ICECAP

****

A Novel of Unexpected Consequences

By

Eric Shulenberger, PhD JD

3912 NE 127th Street, Seattle WA 98125
ericshul@hotmail.com (206-367-5886)

© Eric Shulenberger 2018

Available from AMAZON print-on-demand:
search "books shulenberger"

ICECAP

ISBN-13 = 978-1545485989
ISBN-10= 1445485984

Library of Congress PCN = 2017906075

Produced by Shulenberger Publishing
3912 NE 127th St.,
Seattle WA 98125 USA:

ericshul@hotmail.com

A work of fiction
Copyright © by Eric Shulenberger, 2018
Second printing, with emendations included.

All rights reserved. Except for short passages used in reviews, no material from this book may be used in any manner for any reason or purpose, without express written consent from the author.

Table of Contents

Section One – Greenland
- Chapter 1 – The Ice Cap — 07
- Chapter 2 – In the lab, ashore — 35
- Chapter 3 – Jayhawk imperiled — 45

Section Two – India's "Mean-Sea-Level" People, a.k.a. Just fourteen centimetres from nuclear war…
- Chapter 4 – Ramdaas introduced: his education — 79
- Chapter 5 – "Migration Project" conceived — 92
- Chapter 6 – On PARANOIA – governmental & other — 110
- Chapter 7 – PANIC! When 14cm becomes 40cm — 117
- Chapter 8 – "Intelligence" machinations: India, Pakistan, Bangladesh — 123
- Chapter 9 – GONE MISSING – one atomic bomb, Model N-17, 200 kilotons — 127
- Chapter 10 – "THE MISSION": Gen. Daisycutter, Col. Hakeem, and Nukes on the Mudflats — 132
- Chapter 11 - RESPONSIBILITY – the ultimate mantle — 139
- Chapter 12 - Nuclear denouement – rewards — 151

Section Three – Freshwater vs Gulf Stream
- Chapter 13 – Jayhawk imprisoned — 161
- Chapter 14 – Jayhawk: science plan — 163
- Chapter 15 – Jayhawk: informing the world — 168
- Chapter 16 – London: crafting a response — 197
- Chapter 17 – Jayhawk: Puddle Science — 210
- Chapter 18 – Jayhawk: freedom + some dismal science — 214
- Chapter 19 – SHUTDOWN — 224

Section Four – Antarctica – long-term consequences
- Chapter 20 – Antarctica – The Tiger in the Bamboo — 227
- Chapter 21 – Afterword: London — 232
- Chapter 22 – Loki's second shoe drops — 235

Foreword

I am a retired biological oceanographer (PhD from Scripps Institution of Oceanography [Univ. Cal. San Diego], 1976). My scientific career was spent as a global-scale ecologist, studying the Antarctic Circumpolar Current, the North Pacific Central Gyres (there are really two), and various other things. Like one protagonist in "ICECAP", I am not formally a physical oceanographer, but have learned something about that topic through time and experience – because to understand, say, the North Pacific as an ecosystem absolutely requires a detailed understanding the physical, chemical, and light environments – plus others.

Two underlying premises of this novel are (1) Mother Nature is still very much in charge of human affairs, even though today we humans do meddle successfully around the edges of our homeland Earth (usually with unfortunate results); and (2) various bits and pieces of this planet we humans infest are interconnected in complex and often utterly unexpected, unpredictable ways – with even more wildly unpredictable consequences for various human interests.

In the stories that make up this braided tale there is nothing physically impossible, or even highly unlikely – in fact, over short geological time-spans (but longer than most humans are capable of thinking clearly about - e.g., a few tens of thousands of years), all the major scenarios in ICECAP are pretty much certain to occur… I have just taken the "author's liberty" of having them occur in the present and/or in the very near future.

Hopefully the story (stories) herein will both amuse and educate readers, and perhaps encourage them to take a somewhat different view of our world's complexity and inter-connectedness.

---Eric Shulenberger, Seattle WA, 2018

Other books by Eric Shulenberger

Deny Them the Night Sky – a History of the 548th Night Fighter Squadron. ©**2005**: 520 pages, 998 illustrations and photos in text, plus all illustrations and photos on an included CD. Privately printed, one edition, one printing, 1093 copies.

Forty Thousand Years of Separation. A novel of possibilities. ©**2016:** 289 pages. Available at Amazon print-on-demand (search "Shulenberger books").

CAUTION: Some Parts May be Accurate! An autobiography in random short stories ©**2016:** 419 pages. Available at Amazon print-on-demand (search "Shulenberger books").

Sick and Hurt. Medical Events in O'Brian's Aubrey Novels. ©**2016:** 264 pages. Available at Amazon print-on-demand (search "Shulenberger books").

E'KU – English (Pseudo) Haiku. ©**2016:** 71 pages. Available at Amazon print-on-demand (search "Shulenberger books").

Plus several dozen scientific papers
on oceanographic topics.

About Timelines

This story is complex: some important parts occurred or began years ago (some of those parts are still ongoing). Other parts occur in blindingly-fast present-day sequences, with those sequences scattered amongst larger, slower events. As seems helpful, the text provides relative times for chronicled events, those times being written thusly:

$$\{T_0 +/- \text{ some time units, e.g. days, minutes}\}$$

The core temporal reference point for the stories is "T_0" which should be read as "time zero". This is the moment when, unobserved by humans, the geological trigger was pulled -on Greenland- which started the critical chain of events. An event occurring 23 hours after T_0 would be given as in rocket-launching, namely $\{T_0 + 23 \text{ hours}\}$ - {-, a.k.a. 'minus'} means {before T_0} and {+, a.k.a. 'plus'} means {after T_0}.

SECTION ONE - GREENLAND

1

THE ICECAP

{T_0 – a few days}

• • •

PROLOGUE - GODS PLAYING DICE

The Gods, as usual, sent warnings well before they acted. This time, the exalted get-together was an all-boy event: Poseidon, Vulcan, Oceanus and Neptune (plus a horde of invited lesser figures) were playing dice, betting on outcomes from their invented problem. The warnings arrived unheralded and unlabeled, intended for two groups, one of oceanographers, another of glaciologists, all embarked together on the same ship. Although the warnings were superficially different, in fact both were quite dependent upon the same underlying aspects of the ocean. The Boys had set up the problem and gotten events rolling, then forbidden any further diddling with the problem by any immortal.

The first gambling question posed by the Gods was 'Will the warnings be detected, understood, and properly classified?' Second, if 'yes', then 'Would the Mortal recipients do anything useful with the information?' Betting was running strongly against correct identification, and almost ridiculously against 'doing something useful'. As a result of the gods' manipulations in set-

ting up the problem, a great many interlocked events happened in a very compressed time-frame. There were also the occasional wild long-term gyrations, powered by the laws of unexpected consequences. Those results require that the stories be told in both sequential and time-parallel bits – the reader must not expect a linear narrative when nature herself was not acting linearly.

• • •

INTRODUCTION

In America, congressional influence sometimes gets applied in peculiar ways. The politics behind a national-asset icebreaker being named for a mythical bird representing a thinly-populated inland state that has only fossil oceans (and no permanent ice!) were nonsensical, byzantine, and simply amusing. American icebreaker WAGB-69, formally named "Jayhawk", is a brand-new Healy-class 420-foot, sixteen thousand ton icebreaker, diesel-powered, thirty thousand shaft horsepower. Officially a "medium-weight" icebreaker, she is nonetheless the pride of the US's tiny fleet of icebreaking Coast Guard cutters – the USA's biggest and best, about 3/4 the size of Russia's dozen nuclear-powered behemoths.

Presently about halfway through a month-long cruise, Jayhawk is hove-to some four or five kilometres off the steep, mountainous north-eastern coast of Greenland, on duty to do science. In that region, the edges of Greenland, the world's largest island, are almost un-surveyed – although the latest charts show the mouths of frequent, steep-sided fjords, a mere couple of kilometres inland such geographical data give way to blank white expanses that in earlier times would have been labeled "Here Be Dragons". The unlabeled white area is the Greenland Ice Cap, which covers well over 95% of the island and is topographically boring. That cap, by best estimates (which are none too precise), contains about three million cubic kilometers of fresh water – 7% of Earth's total supply. Enough to raise global sea-level by about 8 metres, were it all to melt.

Jayhawk's scientific contingent is busy doing the research for which they had received both funding and ship-time. In the overall scientific party there are two main groups: the larger is comprised of glaciologists who are installing the last few of a massive network of 500 acoustic-sensor packages, each precisely positioned. The first half of the net was installed last year. The network covers the entire Greenland Ice Cap: the goal is difficult and ambitious - to map the icecap thoroughly in three dimensions, using a specially-designed new type of sonar. This is a very large, expensive and long-term project. The second, much smaller group is of biological oceanographers (cleverly nicknamed [amongst friends] as "BOs") bent on studying high-latitude open-ocean ecology. This is a bit of the ocean that is difficult to get to and seldom studied. The BOs' goal is to make detailed measurements of very basic ecological parameters.

At the moment, both of Jayhawk's helicopters are far out over the Cap, placing the glaciologists' instrument packages. The Principal Investigator (a.k.a. 'PI' or 'chief scientist') of the glaciologists is a senior academic nicknamed "Doc" Adkins: he is presently aboard one chopper, helping install instrument packages on the ice. Well into his sixties, he still regards difficult field-work as his 'drug-of-choice', and would love to have assigned himself as part of one of three teams of graduate students spending almost two weeks on the Cap itself, but resisted that temptation, opting instead to trust in his students to do the work correctly – independence of action being part of their training.

Meanwhile, the ship is station-keeping - remaining in one location - and the BOs are trying to make their measurements.

{T_0 + about half a day}
FIRST WARNING

At the forward port hydrographic winch, senior BO grad student Rebecca Wilson, leader of the BO contingent, was chatting nervously with James, her program's scientific-electronics engineer – a widely-experienced, talented and expensive person, paid for out of her major professor's grants. Rebecca's work involved intensive use of a soccer-ball sized instrument nicknamed "football" and weighing perhaps ten kilos in air - it can be lowered over the side to any depth less than 2000m, to measure several important ecological variables: concentration of dissolved oxygen; amount of chlorophyll in the water (an analog for the amount of photosynthetic organisms [mostly algae] and their activity); water temperature and salinity; depth of the instrument, how much light is present, and others. The whole football system had been designed and built by James.

Rebecca's group was flustered – they'd just encountered what she was certain was an instrument problem. She and her team of four other, more junior students had checked out the football before putting it over the side and into the water, found everything fine - all sensors were "GO", all responded properly to calibration. All batteries were brand-new and had been double-checked. Nobody had admitted to inadvertently bouncing the unit off the steel deck. There was no obvious reason for any of its sensor-systems not to work. But when this expensive, properly calibrated, known-to-be-okay instrument was hooked to the wire and lowered over the side, they got nonsense. Zero salinity. No chlorophyll. No dissolved oxygen, either. But the temperature and depth sensors were fine - they reacted instantly to submersion, gave sensible readings. The group had wetted the instrument, and hauled it back aboard, several times, always with the same results. Rebecca was getting both worried and pissed - her expensive, rare wire-time was evaporating.

James plugged his portable test rig into the instrument and ran a complete automated systems check. The process took about 20 seconds, reported things to be 100% normal. He showed the results to Rebecca, who snorted and said "James, that's EXACTLY what I've been saying... just like we always do, we've gone through the same set of checks in the lab and got the same result as you just did. But out here in the real world we're getting gobbledygook! ZERO salinity? Bull-puckey! We're out here in the middle of the goddamned OCEAN, not a lake of distilled water! It HAS to have salinity greater than zero! Zero chlorophyll? Double-puckey! We're in high latitude at midsummer when there is ALWAYS a bloom of algae. There HAS to be chlorophyll in the water. Zero oxygen in the water? Triple-puckey! Likewise there has to be dissolved oxygen from algal photosynthesis. But not according to this gizmo! Watch!"

Her crew of undergraduate volunteers lowered the instrument by hand to a metre subsurface. James watched the display – the readings were exactly as Rebecca had claimed. Zero oxy, chlor, and salinity; depth and temp just fine. Rebecca faced him, hands on hips, upset showing. James mulled things over silently for a moment, then shrugged and said "So, out here on deck our device seems to be ops normal and calibrates just fine. But today, being wet it just doesn't like. This is a damn good instrument, never given us any such problems before. Weird! Maybe Murphy is on board Jayhawk."

He paused, grinned, scanned the undergraduates' faces. Frowns of puzzlement all around. "The reference, folks, is to Murphy's Law, which correctly says 'Anything that can go wrong, will.' Back in 1979, when none of you were born yet, there was a very famous accident at the "Three Mile Island" nuclear power plant. One of the big reactors ran away and its core melted. Worst nuclear accident before Chernobyl. You've probably all at least heard of it, right?" This time, nods all around. "OKAY, so tell me - does anyone in this group know WHY the core meltdown happened?"

Nothing but negative head-shakes: Rebecca started to say something, James stopped her: "The reactor core melted because the operators either couldn't physically SEE their instruments, or refused to believe them."

Then, sensing a great teaching moment, he went on; "I long ago learned to always start out by believing my instruments. This here football is a very good and reliable instrument. Until convinced otherwise, I choose to believe it. We need more information. Let's lower the unit down to say fifty metres and see what that gives us."

Two male-student strong-bodies hand-over-handed it - the others all watched the display. At 17 metres, almost with an audible click, salinity jumped from zero up to 27 parts-per-thousand, appropriate for this area and season. Simultaneously, chlorophyll went from zero to a reasonable value. As did oxygen, the byproduct of chlorophyll-based photosynthesis. All three parameters now read out as perfectly normal.

Dead silence. James grinned delightedly, then asked loudly, "HEY! Has anyone thought to TASTE the surface water? Human senses are pretty reliable most of the time. For wild-and-crazy data like these, some corroboration would be good!"

Even today, high-tech everything aside, when one needs a surface sample of the ocean, a bucket on a rope is far and away the best and easiest tool – not to mention cheapest. Just don't use the bucket for anything else... ever! Rebecca turned to a junior trooper: "Kelly, please go get the surface bucket." Under her breath she muttered "Fucking goddamned idiocy anyhow! What the hell is going on? None of this computes in any way!"

The special "uncontaminated seawater only" plastic bucket arrived, went over the side, returned full.

"Well?" James asked bucket-handler Kelly - "Aren't you going to do the science, man?!"

A bit of a closet grand-stander, Kelly shrugged, dipped a finger, tasted, looked surprised, then raised the bucket to his lips and took a long, deep swallow. Up came his face, wearing a silly smile. He delivered his judgment: "Ice-cold, and fresh as the day is long. Folks, we do be floating in **SALT-FREE** water, not just low-salt." He grinned again, held out the bucket: "Hey! I thought I was going on an oceanographic junket – am I being cheated?"

One by one they all tasted. Perfect consensus. Fine, eminently drinkable freshwater. Offshore from Greenland, and according to their single instrument-lowering, the ship was presently floating in a 17-m-thick layer of freshwater. Improbable, but obviously not impossible. 'Believe your instruments!' was duly internalized by all.

James was a good oceanographer himself, after more than thirty years at sea helping every conceivable variety of oceanographic PhD-holder and wannabee. He put on a lecturer-face to which he was fully entitled, addressed the BO group: "Now, folks, I personally don't have a clue what's going on in this part of the ocean, but dammit, people, you gotta BELIEVE your instruments... otherwise you'll fly right into the mountain. Never forget that! Rebecca, when you-all figure out what the hell is going on here, let me know. This ship appears to be in the middle of the ocean but somehow we're floating in a puddle of distilled water that has in it no biology and no oxygen. Bloody funny business, seems to me! Maybe the folks on the bridge would like to know about this, don't you think? And somebody is going to have to write a journal article about it! Better all of you should keep good notes, for sure! How else can you become rich and famous?" He turned and left, carrying his test rig.

SECOND WARNING

Meanwhile, in parallel with James and Rebecca's "football problems", there was occurring on the bridge the second warning. It involved station-keeping ... holding the ship in one place, as required by many research projects, including today's BO work.

An hour or so ago, beginning just about the time the BO-football was being prepared for launch, Lieutenant Julie Jonson, the duty watch-officer on the bridge (meaning standing in the shoes of the Captain), encountered a small problem maintaining proper position of the ship. She was having to manually reposition Jayhawk every few minutes, using the GPS - an irksome and thankless task supposedly the responsibility of a pair of electronic systems.

The Lieutenant was presently complaining to the bridge's electronics tech about the navigation gear. Pointing to Jayhawk's plotted track for the past hour, she exclaimed "Petty Officer Schultz, just look at this plot! It looks like we've been doing some sort of cha-cha! Our million-dollar integrated digital navigation system is in charge, and this is what we get? Baloney, PO Shultz!"

Curious and puzzled, they called up and displayed the last several hours' navigation data.

After studying the plot, with its unusual jigs and jogs, Jonson asked again, "What the dickens is going on here, anyhow? Weird! Any thoughts, PO?"

He replied "Well, Ma'am, it sure looks like a big part of the problem is that we're drifting south at about a knot. I didn't think there's supposed to be that much current around here, but there you go. To counteract that drift, every so often the GPS part of the system kicks in a little rudder and main-engine power - but only intermittently, and not a great deal of it. That plus your own adjustments account for the major wiggles in this plot."

"Hmmmm" said Jonson. "So the first problem is that we're unexpectedly drifting south. Peculiar. You're right about there supposedly being very little current here, this far north. But there's supposed to be a weak near-coast "East Greenland Current" going south. That's what the charts say, anyhow!" Then she laughed: "Of course, the charts are probably based on one observation made by the drunk captain of a fog-bound square-rigged whaler back in about 1827 or so. We are NOT in a busy part of the world ocean! What we see and experience for ourselves gives us better info than we can find on any charts, you betcha!"

She scanned the plot again, muttered "Well, there's nothing dangerous or unusual south of us at the moment, so at least the current's not setting us into something undesirable. The GPS seems to be handling the overall drift ok, with some help from me." She sighed: "Nice for a mere human to still be needed in this 'computer-and-satellite' day and age! Odd, though, us being out here in the open ocean and leaving a wake because we have to make turns in order to stay put! So… any specific implications for the nav system, PO?"

He cocked his head, thinking. "To me, Ma'am, this plot means the GPS half of the system is working okay."

Jonson nodded agreement, then tapped the plot with a fingertip. "But John, something's missing here. Our fancy nav program usually calls for lots of little-bitty fine-scale corrections, using signals about ship drift from the Doppler acoustics to drive the positioning thrusters. Sometimes twenty or even fifty corrections in any given minute. Those corrections are pretty trivial compared to the big moves called out by the GPS – they just show up as a fuzzy fringe of little corrections scattered around the main track. They're not here! Gone missing, they have! Just look!" She enlarged a section of the plot.

PO Schultz again nodded agreement, said "You're right, the Doppler stuff is missing for sure. Probably something electronic gone wrong with the unit. Which certainly makes all this into my problem, doesn't it, Lieutenant? As 'troubleshooting step one', let's wind this plot back in time until we see where the fuzz disappeared. Maybe something happened at that moment, and knowing about it might just help me fix this sick puppy."

The course plot back some few hours contained the break: fuzzy, then no fuzz for a while, then the fuzz returned briefly, disappeared again and remained gone thereafter. John pointed, muttered "Well, shoot, Ma'am. Something happened right here. Either the transmitter went down, or the receiver failed. I need to go do a bunch of signal-checking at the gear itself... give me a few minutes, Lieutenant, and I should be able to figure out what's gone wrong. Ten more and I should have it fixed and ops normal again... it's not a very complex item – no moving parts at all!."

"Please make it so, PO."

PO Schultz disappeared, returned in half an hour looking distinctly puzzled. "Well, Ma'am... the transmitter is going bang bang bang just like it's supposed to... frequency, power, pulse repetition rate are all exactly right. The receiver works just fine when I put it on the bench, but back in the unit and into the water, it either isn't getting a signal, or it's losing it after it comes in. Really whacko – I even swapped in the spare – got the same results."

He smiled, shrugged, said "Then, just for fun, Ma'am, I clipped a set of old-fashioned earphones into the receiver... you can hear the outgoing bang just fine, which means both the transmitter and the receiver are working okay. But I don't hear an echo. Weird. Big-time weird, Lieutenant."

Lt Jonson contemplated things, not wanting to make some sort of dumb-bell remark. A quick mental review of what she could recall about the Doppler system. PO Schultz waited patiently, silently – he'd long ago learnt that it never paid positive dividends to interrupt a thinking commissioned officer. Finally she said "John, if I'm off base too far, please correct me. The outgoing sound has to bounce back, return to the receiver as an echo, right? An echo off some sort of solid target." He nodded – so far so good. She continued; "And the sound won't bounce back off the water itself... instead it bounces off little animals, off of zooplankton, correct?"

"Yes, Ma'am, exactly so. The zooplankton is a critical part of the system, but we don't have to supply it, the ocean does that for us. Lots and lots of little targets... tens of thousands per cubic meter, so the biologists tell me. Everywhere in the ocean, all the time. At least, anywhere near the surface."

The Lieutenant nodded intelligently, said "So –even if all the electronic subunits of the Doppler gear are working just fine, but there is no echo, the system is functionally dead, correct?"

"Yes indeed, Ma'am."

"This sounds rather like we're in a piece of the ocean where there happen to be no zooplankton in the water. Or have I drawn a stupid conclusion?"

"I had the same idea and got the same result, Ma'am ... but I've never, EVER heard of a place in the ocean where there isn't lots of zooplankton. Maybe we should ask the BOs about that?"

While the pair were contemplating the situation, Captain Shelton arrived on a walkabout, was offered the bridge and courteously refused it: "You keep the watch, Lieutenant. Anything interesting going on?"

Lt Jonson recapped the positioning conversation. As she finished, and before the Captain could respond, Rebecca arrived with a liter jar of the bucket-sample water. The Captain raised one quizzical eyebrow at Rebecca, who got the hint: her turn. "Captain, we've got something really weird going on. The surface water we're sitting in has NO salt – ZERO, from the surface down to about 17 m. That water also has no biology, and no oxygen whatever. It's not an instrument SNAFU, we're sure of that. It's plain crazy, Captain ... no such conditions have EVER been found anywhere in the ocean! For sure, not much ocean life other than birds and mammals can survive in zero-oxygen water!"

The Captain looked exceedingly skeptical: Rebecca soldiered on, held out the jar. "Here --- taste it for yourselves, this is five-minute-old surface water."

The officers eyed one another, then each tasted, expressed astonishment. Rebecca continued – "We lowered the instrument by hand and at 17 metres everything went back to normal. Blink – just like that! Salinity 27, lots of chlorophyll, lots of oxygen. I think we're floating in a really BIG puddle of freshwater. But not just any old freshwater, this stuff is biologically sterile! I have no clue what-ever as to what's going on here." She finished with "James insists that we believe our instruments, and he's right, of course. So where do we go in order to find enough freshwater to make a 20-metre-thick puddle out here? It must be from Greenland, I believe, but how the devil!?"

Nobody said anything for some seconds, until PO Schultz broke silence. "No chlorophyll? That means no algae. And no plants means no oxygen. Which means no zooplankton. I think we've probably identified why the ship's Doppler positioning isn't work-ing right. The sonar transducer is sitting in the upper layer, working perfectly but without any zooplankton to give an echo! I'll bet you all that the break in the Doppler data happened when we ran into The Puddle of freshwater. Actually, because we are station-keeping, it's more correct to say that The Puddle drifted down from the north and ran into Jayhawk, not the other way." He looked at Lt Jonson and said "Looks like we figured things out just about right. Maybe, Ma'am, you should become a technical troubleshooter – you're off to a good start!"

The Captain said to the group, "People, if this screwy surface layer, this puddle, is thicker than our draft and is also going south at over a knot, that is one hell of a lot of freshwater, Like Rebecca just said, the first question is, 'Wherefrom cometh the water?' Rebecca's right, I think – it has to be coming from Greenland somehow. The quantity is huge! Plus there's the question of where the energy comes from to push it along? He scratched his head: "Not that I don't trust your data, everyone, but let's get some more information. If we're somehow afloat in freshwater, then we have to have gained at least a foot of draft." He then turned to the bridge's duty seaman: "Go look at the Plimsoll mark... you'll have to lean way outboard to see it, don't go falling overboard. Get someone to help you. Report back here with your findings asap."

Ten minutes later they got confirmation – Jayhawk was, indeed, riding almost 35 cm deeper than she had been earlier: she was now sitting almost exactly at the "FW" (i.e., 'fresh water') line – a mark normally useful only when going through the Panama Canal or navigating a major river like the Mississippi or the Ganga.

"Well..." said the Captain, "...we're supposed to be doing science, discovering stuff, so Rebecca, you're the duty oceanographer aboard even if you're of the "BO" subspecies. I doubt our resident glacier group has much background in wet oceanography ... in fact, it's likely you've had a lot more physical oceanography training than the rest of the crew and scientific party combined. So getting an explanation for this falls squarely into your bailiwick. Therefore you better round up your herd and puzzle out where this much freshwater is coming from! It must mean SOMETHING!" He paused, laughed, pointed to Greenland's eastern-edge mountain range, to port in the near-distance: "There's a world of frozen freshwater over there on the other side of the mountains. Whole darned ice-cap, three kilometres thick and the area of a small continent – and it's 100% freshwater. Seems to me that to explain our situation all we need is an energy source to melt some of that ice, plus a mechanism to move it, plus a handy large-diameter hole in the mountains for it to flow through, and voila! We explain our local mysteries and you guys have some great new science to report! Or maybe it'd be more like science fiction, or even science fantasy?"

• • •

The Gods in their gaming-hall pricked up their ears – the Mortals seemed to be on the verge of understanding. But it didn't quite turn out that way.

• • •

The Captain was correct – Rebecca had by far the best background for studying this huge puddle of freshwater, to describe it and then try to fit it into our overall knowledge of the ocean. Rebecca took her informal, Captain-created position as 'ship's physical oceanographer' seriously, and she knew a potentially career-making discovery when she saw it. She knew instinctively that the freshwater had to be connected to Greenland, and more specifically to the island's ice cap - there was no other conceivable source.

Ordinarily, Rebecca's first instinct would have been to call for satellite imagery of the area, to check the size and rate of increase of The Puddle… and perhaps identify the source as well. But for the past forty-plus hours, since well before the onset of the freshwater weirdness, high clouds had made satellite photography useless – both for ordinary imaging at visible wavelengths, and for infrared images of sea-surface temperature.

Frustrated by the clouds, she decided to bring all the BOs together at mid-afternoon to discuss the phenomenon. The glaciologists were invited, too, but only a couple attended –at the moment most of them were preoccupied with their own group's work, setting out the instrument network. In addition, most admitted quite frankly that they weren't much interested in liquid water of any sort - freshwater in the ocean, for them, was a big 'So what?' – of no interest unless the water in question happened to include the big temporary lakes of melt-water running about loose on the GIC, doing unknown but surely evil things to the Cap - a very recent phenomenon, due mostly to global warming.

The group convened in the library. Rebecca recapped the observations, and then announced, "The freshwater has to be coming from the GIC – that's a given...unless someone wants to argue?" There were no contrarians. "When the weather clears, we should be able to see the origin and size of The Puddle, but not yet. So we get to brainstorm. It's a difficult problem - even given the source, there's no obvious mechanism to get this much water either through or around the east-coastal mountains. The floor is wide open for speculation – but invoke me no aliens, please. All other apparently-nutty ideas welcome."

Nobody else volunteered to go first, so Rebecca did: "I'll start. Let's begin by following the Captain's lead – he rattled off a good little list of things we need to solve this mystery. First, a source for freshwater. The water we have available is solid ice – the GIC. So if we are going to use that source, we need a humongous source of energy to phase-change the ice into liquid. Then we need a hole in the mountain for the water to flow through – unless it is somehow coming around the north end of Greenland, which seems very unlikely. Plus energy to shove the freshwater through the hole. Finally, we need a power source to drive the resulting puddle south faster than the underlying oceanic current. Lots of stuff we need to know!"

"Now, back to our list – we've located the needed water but it's frozen. So, group, where do we get enough heat to melt this much of the GIC? Any good ideas? Or even bad ones as starters?"

George, a final-year glaciology student, was off-duty and had decided to attend. Now he joined in: "I'm a geologist more than a glaciologist..." he said. "So I have a different view from you oceanographers. If it were up to me, I'd begin by observing that whatever is giving us this puddle, it is a SUDDEN-ONSET phenomenon. The puddle wasn't with us yesterday. Sudden onset of The Puddle suggests to me a sudden input of energy. The energy has to be sufficient to melt oh, say for fun a couple of cubic kilometres of ice-cap pretty much overnight. That's a LOT of energy, folks! We're talking geological-scale energetics. Disregarding the "heat source problem", if it's going to melt ice, the necessary energy, specifically as heat, has to get TO the ice cap somehow, from wherever the heat was generated. How can it get there? It has to come in from above, from below, or sideways – those are the only three possibilities. Unless the heat were somehow generated in-situ – and since we're dealing with a small continent buried under ice three kilometres thick, that sounds unlikely. We can consider the three possible paths in turn. From the top is obviously nonsensical – there's not much solar heat energy getting to the surface at these latitudes… and besides, this puddle is a sudden phenomenon, and so far as I know there's been no sudden influx of overhead heat –meaning solar input - to this part of the world. We'd have noticed a solar flare that could do this! Coming in from the side is equally unlikely – the island is fully surrounded by icewater ocean, and most of the GIC is well above sea-level too --- which means that if you brought in the energy horizontally, say as a warm current, you'd have to shove that water uphill a long ways before you could even begin to melt the ice you'd need – and it wouldn't do the melting overnight, like this phenom calls for. I'm stuck. Sorry."

Kelly from the BO group piped up: "I like the way you're going, George. It seems to me there's a workable, sensible alternative on the heat-supply problem. Rebecca, you called for wild and crazy, so here's mine. What we have for sure is a huge RELEASE of energy, and it's got to be a pretty FAST release. But that doesn't require a sudden INPUT, or even a huge input, of energy… not if you could STORE UP the energy, say like a battery on a trickle charger, and then release it all at once. Lots of human and natural systems work that way - think race-riots, earthquakes, volcanic eruptions. Maybe something, some process, managed to store energy, accumulated a lot of it, and then along came a trigger and set the system into motion? That scenario wouldn't require some ungodly improbable huge burst of energy input yesterday in order to have today's phenom. The system could have been winding itself up for millennia!"

"Great idea!" said George, letting show a bit of excitement. Clearly he had thought of something else. "I've looked at the geology of the island itself – most of the others in my party are interested in the ice, but I'm fascinated by the rock in the island's root, sort of down in the basement. We don't know much about the ice-cap, certainly not in 3-D – but we know even less about the big rock cup the

GIC is sitting in – that's my niche. Nevertheless, we DO know a little, and it might be useful. In particular, folks, we know there's a hot plume of mantle magma under the center of the island. You can see it clearly in gravity measurements made by satellite. If that plume just sits there for a good geological while, heat from it will eventually work its way up to the interface of ice cap with basement rock. An energy-storage mechanism is what we really need, because I don't believe we're going to come up with a sudden massive blast of ice-melting energy. And yes, there's a nifty mechanism available for storing the energy. The heat is moving, by conduction and advection, right through the rocks between the GIC and the top of the plume. Now, rock is a great thermal insulator, so you can pump heat upwards only very slowly. Slow doesn't matter – long-time forcing DOES matter. Any heat that fights its way UP through the rock encounters ICE three kilometres thick, and ice is an equally good thermal insulator."

His excitement broke through: "Think about it, folks! Here's your storage system. Heat from the plume moves up - a very slow flow due to rock's insulating properties. Then, just when it finally gets to the upper surface of the basement rocks, it encounters an obstruction made of another insulator."

He paused to survey his audience, which was definitely paying attention. "Now we have an interesting situation. The heat coming into the interface can't go DOWN against the gradient from the plume, and it also cannot go UP through the insulating ice. The rising heat gets TRAPPED, and over geological time frames if you keep pumping heat into the rock+GIC system right at the interface, and there is no effective heat-drain, then the interface –which is ice– will get warmer and eventually it'll MELT. Then, at least we'd have all the stored energy we need to move lots of water around! Both thermal and gravitational potential energy. Importantly, visualize your glass coffeepot in action, but on a global scale. Water is heated from below and expands, becomes less dense, rises… carrying heat. In our large-scale scenario, carrying it right up against the GIC's bottom. Over geological time, that HAS to be incredibly destabilizing for the Cap!"

Then George told them, "I'm sort of a collector of geological catastrophes. Want a scary scenario? Plain old liquid water is a fine lubricant. What we're talking about here is a possible huge – really HUGE!- lake of relatively warm liquid water underlying a couple of kilometres of ice, all at high altitude. The GIC is enormous, well over three times the area of Texas. If ever the GIC ice started moving, it'd be lubed by the lake water, be powered by gravity, and I cannot imagine anything that could slow it down, much less stop it. The whole damned Cap could just slough off and run downhill into the sea. All three million cubic kilometres of it." He shrugged, smiled, and asked, "Scary enough for everyone?"

He stopped for a breather, looked about for questions, then continued. "What would happen depends critically on the internal structure of the actual ice itself, however thick it may be. The exact 3-D structure of the GIC is what we're trying to measure with those 500 sensor packages we've spent two years covering the GIC with. The ice is complex and VERY difficult to see through using acoustics. We really do not know some very simple stuff… such as, exactly how thick is the GIC? We all bandy around this "three kilometres" thickness but we don't really know that number at all accurately. How about other stuff we don't know? Try these questions: 'Is the GIC solid ice from top to bottom?' or 'Is the bottom of the cap actually attached to the underlying basement rock?', or 'Is there a layer of liquid water at the rock-ice interface, just like what we're talking about here?' Nobody has decent data, measuring anything through a couple thousand metres of ice is a royal bitch of a job. You can't drill a deep hole for sensors because the ice is plastic and being under pressure it fills your bore-hole right behind the bit. You can never even get a drill-bit back!"

He chuckled: "In biology, that would be like the bacteria always eating the Petri dish, or something. People have tried, though. We're here doing this project because decades ago some scientists set off a few 5 kg blocks of TNT, up on top of the Cap, and recorded the echoes from the rock bottom. Primitive, but effective. Unfortunately there are two ways of interpreting the data – one group says "Ice all the way to the rocky bottom!" and the other group says "Big lake of melted water between basement rock and underside of Cap." Exact same data yield two very different scenarios. If it's a LAKE down there, then there's all that liquid water – God only knows how much of it- and it's being held back by an ice dam of some sort. All that water –and ice- is sitting at a mile or two of altitude and wants oh-so-badly to go charging off downhill at the slightest provocation. We don't have a real clue as to which it is, lake vs attached ice. That's probably the biggest question we glaciologists are trying to answer. Heck, maybe both ideas are wrong and some other screw-ball hypothesis is correct. On the bright side, we're going to get a beautiful 3-D picture of the whole GIC using our instruments – in fact, sometime shortly we ought to be getting a first rough picture of the half of the cap that we instrumented last year – those 250 units have been sending basic info back for a year now."

George quit talking. The whole group sat stunned for a moment, until Rebecca said "So we get the energy stored, and it's in the form of melted ice, warmed-up water, at high altitude. Seems we have both plenty of potential energy and the necessary water supply." She looked at George for confirmation as she said "Now let's think about the hole, the passageway. I'm surely no geologist, so correct me if I screw it up."

"That plume from the mantle means the island is atop seismically active materials: a seismic event, meaning earthquake, maybe even just a small one, might create a crack, or open an existing crack, right through our mountains out there. Who knows how long and how wide a crack you'd need in order to move that much freshwater!? And the power to move it through a crack... we still haven't got that! How are we going to exchange some of the gravitational potential energy for energy in a form that can shoot water through a long skinny tube?"

George thought hard for some seconds, then said "OK, here goes another wild one. Recap - first we need a crack from under the GIC extending all the way across the roots of the coastal mountains – a connection between the open ocean and whatever is under the GIC. Seismicity and Dr Murphy could generate that connection. If we assume a crack, then we need to push water through it. A hell of a lot of water. In our brainstorming so far, I think the crack is the weak point –so to speak, no pun intended. But IF we had an opening, there's plenty of energy available to drive water through it. Gravity acting on mass, uber alles! The energy we need is stored as the gravitational potential energy of our unseen hypothetical high-altitude lake, and that form of stored energy can be put to use hydraulically. It can easily be turned into the kinetic energy of flowing materials."

Geologist George grinned, obviously enjoying his unexpected, sudden occupancy of the limelight. He was on a roll. "Now, about the hydraulics – let's think first about the water out here in the ocean, at a depth of say 3000 m below mean sea level. The static pressure at that depth is pretty high, about 300 atmospheres a.k.a. 4500 PSI. But now think about the water under the ice-cap, at the same depth below the surface, namely 3000 m. That water has the added pressure of three more kilometres of water atop it. So the under-ice water at 3000 m depth has a static pressure of 6000 m of water, or about 300 plus 300 atmospheres... say well over 9,000 PSI. For sure there's a huge pressure difference between free seawater and pressure at the bottom of any trapped liquid. The difference is at least 4500 PSI. That's seriously high pressure, folks. Put the required crack in the mountains so that it connects the two very different pressures at 3000 m depth – namely outside and inside... and BINGO, you'll get flow. Probably a LOT of flow... after all, we are talking geological scales here, not automotive hydraulics! The flow will be from high to low pressure, of course, which is what we need... fresh-water flow from under the ice cap out into the open ocean. The pressure difference extends over a huge area and consequently it could drive a big flow of fluid. Really BIG!"

A freshman BO graduate student tentatively lifted a finger off the table, signaling that she wanted to say something. Rebecca nodded at her to go ahead. "Just a small crack through the mountains doesn't sound adequate. No matter how much pressure you put on it, not much flow is going to happen. We have to get a much larger hole, not just a crack. I think I can help." The others waited expectantly. "My oldest brother works for a steel mill. Two years ago they got a new steel-cutting machine. It uses ultra-high-pressure water to cut intricate patterns through steel plates up to several inches thick. No blades, no diamonds, no flames. Just water at crazy pressures. He took me to see it – it's AWESOME, no noise, no heat, no metal sawdust or grindings. Just this mysterious-looking sparkling vertical line of water. As a demo, he took a brick and waved it through the stream, and the brick fell apart, the cut surfaces were smooth and shiny. They looked like they'd been glazed! Just from a stream of water!" She took a deep breath. "Maybe if you just got a tiny bit of very high pressure flow through our hypothetical crack, the flow could enlarge its own hole? Maybe the seismic event doesn't need to make very much of a hole after all!? Or is that just too stupid for words?"

George reassured her - "Not stupid at all – you're bang on point. If the fluid contains solid particles it will for sure enlarge the passageway it's squirting through. Streams of water that are moving fast and carrying any sort of particles can be enormously abrasive and they can grow rapidly – often exponentially. Leaks through earthen dams do just that, and the process can lead to catastrophic failure. Quite a few such dams have failed in the USA, lots more of them elsewhere. The dammed-up water provides the high-pressure source. Almost always the collapse is due to some version of what we're talking about."

"There are also such things as naturally-occurring ice-dams. They can fail in exactly the same way. The folks who support the "Lake" theory for the GIC have to postulate SOMETHING that holds back their hypothesized interfacial meltwater, which is at high altitude and on a slope. And it has to be a strong "something' because it has to hold against the pressures of a couple of miles of ice or water. Those "Lake" folks are all totally sold on the idea of an ice-dam… but for such a structure we have exactly zero observational evidence. Which is one reason the "frozen solid" people think the Interfacial Lake people are nuts."

Kelly asked, "Which theory do you believe, George?"

"Myself, I think the evidence is mostly in favor of solidly-frozen and attached to bedrock. Not 100%, but close. Ice dams scare the willies out of me. They can and do collapse just like earthen dams once they begin to leak. We've seen such a thing recently in the USA's Pacific NorthWest. At the end of the Pleistocene, a big ice dam in Oregon and Washington was holding back about 2,000 cubic kilometres of water, forming a huge lake known as Glacial Lake Missoula. That dam collapsed and in one day the lake drained to the Pacific, which was about one good US state's width distant. The flow cut the channel for today's Columbia River – 800 kilometres long, up to 400 m deep and several kilometres wide. In less than one day. Scary. If the Lake theory is right for the GIC, the 40,000 people living on Greenland are on borrowed time, because they are all situated on the lowest pieces of the coastline. Same thing could happen there, but a lot worse... Greenland has 1500 times the water, and because the ice is mostly at high altitude, it has tens of thousands of times the potential energy of the Missoula event. Now THAT ought to be scary enough! In fact, it seems to me that it would be so easy to breach the hypothesized ice dam, that the fact it HASN'T happened, is great evidence that there ISN'T a lake. If the lake existed, it should have gotten loose by now."

He shrugged. "Whatever is providing this freshwater that Jayhawk is floating in, I sure do hope it's not the initial leak in the Lake Theory folk's hypothesized dam! Really, folks, it's not unreasonable to think that IF the Lake Theory is correct, you could send the entire GIC downhill and into the sea – water to begin with, then augmented by the collapsing ice. What it would do to the people living on the Greenland coast, I don't need to think about. Hope I'm correct about 'no dammed lake'! But I certainly won't be buying Greenland real estate any time soon!"

That comment pretty much ended the meeting, which broke up without even trying to reach a conclusion. After the meeting, Rebecca went topside, then up to the bridge. She could feel the engines running, sense the propellers turning over lazily, doing the work of station-keeping. She spoke to the duty officer: "Sounds like we're using main propulsion continually now – or am I imagining things? Earlier it was intermittent."

The officer sipped his coffee, replied "Yep. They're running steadily now, at a fast idle. Your freshwater puddle certainly isn't stationary, Miss Rebecca! I think it's gaining speed, but slowly. We're now making turns for almost two knots, and going nowhere fast. Weird!"

She thanked him for the information, stepped out onto the open wing of the bridge. Weather –specifically meaning visibility- had improved greatly. Looking horizontally she could detect no fog or mist, and the sky looked clear. "Great!" she said to herself, "…time to try to get some reasonably recent on-file sat photos."

Rebecca quickly downloaded several images to use in a descriptive after-dinner talk for all interested hands. She got a good turnout, and the short talk went well – lot of people, both scientificos and crew, were interested in both the physics, and the lack of biology in the freshwater. Discussion was wide-ranging, but mostly expressing puzzlement as to what was going on: no new ideas surfaced. Moreover nobody, including Captain Sheldon, saw anything dangerous about the entire event – interesting yes, threatening no. A two-knot surface current was nothing to Jayhawk, just a little extra energy expended keeping station and an oddity to send off, eventually, to the chart-makers. The freshwater flow -however unusual- would not harm the ship and needn't affect the scientific efforts, it just served as something unusual to amuse the scientificos. No reason to change anyone's operational plans, which were going so spectacularly well.

Consensus, voiced by the Captain: Rebecca et al. should keep thinking about this event, develop a plan to study it after the glaciologists finished deploying their sensor packages. Until then, the ship would continue on its primary glaciological mission without pause, everyone had an intense desire to finish ahead of time, and 100% success was still possible, which performance would accrue kudos to all hands.

At the moment it looked like they would end up with about three days of uncommitted shiptime, which Rebecca could have to do what she thought best in the way of studies. She was stoked – a brand new and obviously important phenom, plus adequate time and resources to study it – not to mention having the incredible luck to be the only qualified scientist on-site at the onset! No competition! She began mentally drawing up the research questions and a study plan.

• • •

The Gods' first two 'early warnings' had arrived at the ship on schedule. And now it remained to be seen whether they would be identified and understood in time to be useful to the recipients. If the information was not correctly interpreted, and quickly, some tens of thousands of humans were bound to die as a result – die together and very suddenly. However, IF the information were understood and acted upon – again, quickly! – those Mortals would be allowed to carry on unharmed. But the Gods took a very Zen approach to the possible slaughter – like today's teacup which is already broken in the long scheme of things, all those Mortals were already dead, again in the grand scheme --- "Death" being, after all, a defining property of "mortal". Not something over which gods should be at all concerned. They themselves being well beyond the bounds of mankind's normal science, Poseidon, Oceanus, Vulcan and Neptune had no reason to be vengeful or antagonistic towards humanity. Fun, not some other, darker motive, was this event's driving rationale.

Now the Gods issued the third and most obvious warning, and with that came a doubling of the flow of mead.

• • •

THIRD WARNING

Well after her post-dinner lecture, Rebecca climbed up to the radio-room, to check whether viewing conditions would allow the satellites to get new pictures of the sea surface. To her chagrin, the high clouds had reappeared, and the surface in Jayhawk's vicinity – out to a couple of hundred kilometres- was invisible to all satellites.

Whilst Rebecca was pouting slightly about the unfairness of it all, George arrived, coffee mug in hand.

"Time for me to check the glaciology program's official ship's email... I handle it for the group, which keeps another layer of crap off Doc's back. Mostly it's data, drafts of papers, et cetera... stuff nobody would really want cluttering up their personal email." He plonked down in one of the radio-operators' chairs, brought up the account.

Rebecca stood idly by, waiting for him to finish: more conversation would be nice, George had been a great help during the brainstorming, and she thanked him again.

"Huh!" he exclaimed, preoccupied... "You're welcome, just remember what you paid for the help! Sometimes cost and value are tightly correlated, say zero to zero. Anyhow, thanks for including me. Fun, doing all that hypothesizing in a vacuum, even if we did just end hanging in midair. Probably lots of other possible explanations – we just got ourselves into the first available rut." He motioned her to look at the screen. On display was a plain, short text message from someone nicknamed Dot Calm.

George explained: "Dot is back at the main lab – she's in charge of monitoring everything. She's the most senior of Doc's students. The best computer jock in our lab, too – she's unflappable, hence the nickname." Then "Read the message... it should interest you after today's brainstorming."

It read:

> **"Hey George! I got the main analytic program up and running earlier today. FINALLY! I tested it on the first few days of data from last-year's sensors, which seem to be working even better than we'd hoped. I managed to do a first-cut analysis –just for fun, not serious science yet. But – just between you and me - I think the Lake people are going to win the interface argument... I think you can really see the end of ice and beginning of water - even with just this preliminary analysis it shows up. Looks like somewhere between 500 and 1000 m of water in the one place I could do**

the analysis. Just thought I'd let you know... pass it along, I'll run the analysis again in a couple of days when we have a better handle on things and can include the whole array. Busy now. Don't do anything I wouldn't do! Hello to everyone. I'll plan to call you sometime tomorrow, say at 2000 Zulu. Be there! BYE!

Unknown to all, the "Lake" theorists are indeed correct. The ice cover floating atop the freshwater lake is structurally unsound, riddled throughout with interconnected holes caused by the trapped interface-lake-water and by melt-water drainage from the Cap's surface. The entire GIC is largely afloat rather than being anchored to the rocky basin: it depends on the underlying lake for most of its support, for its structural integrity.

{T_0 + 60 minutes}

ONSET - THE BOIL'S FIRST HOUR

Unnoticed by anyone, a minor shudder in the basement-basin rock had occurred quite recently. It registered a whole 4.2 on the Richter scale. Ordinarily inconsequential, this time the movement opened a long-developing fracture in the rock and ice down in the base of Greenland's NE coastal mountains.

George's analysis of the hydraulic situation had been bang on. Once fluids begin to move at those pressures, and in so doing begin to carry bits of hard material (basalt, ice), they can cut even hard rock (much less mere ice) like a red-hot buzz-saw. Once a stream begins breaking loose chips of rock or ice, it becomes a self-perpetuating high-speed abrasive jet, cutting through anything in its way, and growing exponentially.

Shortly after that trivial earthquake, the leak started, down at about 3000m below Mean Sea Level (= "MSL"). Within a few minutes, the resulting flow began to entrain particles of ice, then flakes and chunks of rock from the developing tunnel's inner surface. In mere minutes the leak had grown to be a metre-wide, high-speed cutting jet of slightly warm freshwater, being released into seawater at 3000m depth. When it emerged from its expanding conduit, the jet was driven to move vertically, powered simply by buoyancy alone – i.e., by the large density difference between the freshwater and its surrounding seawater. The jet headed straight for the surface far above, gaining speed as it cleared its own path through salt-water. All by itself, that initial buoyancy would have brought the jet to the surface at a seemingly unreal speed, but the freshwater's buoyancy had another, hidden component. Water is NOT incompressible. Reduce the pressure on a parcel of compressed deep water and the parcel will expand, making it less dense, hence more buoyant: that parcel therefore rises, which reduces the pressure on it, and the parcel expands even more - a positive feedback system with exponential consequences. Combined, the two mechanisms (simple buoyancy and expansion buoyancy) drove the jet upwards at an astounding speed.

The cut made by the flow began to affect the underside of the GIC itself within an hour, or at most two, of initiation of flow. Five minutes thereafter the first bits of ice surfaced in the jet, and the GIC's internal structure began to groan and shiver. Ten more minutes and at the surface the jet had become a half-km-wide fountain, to be nicknamed "The Boil". The resulting huge and growing puddle of floating freshwater was spreading rapidly in all directions under the impetus from the supply: the whole puddle was also going south, carried by the underlying East Greenland Current, and pushed urgently from behind by Boil water. Within a few hours the southbound and increasingly thick freshwater flow from the Boil would encounter Jayhawk, there to be discovered by Rebecca's group. The freshwater's effect on the ship's immediate operations would be inconsequential because it arrived so stealthily... the leading edge of The Puddle was only millimeters thick and growing in depth very gradually.

{T_0 + several hours}
COLLAPSE: GRAVITY WINS

On the inland side of the coastal mountains, the Cap's basement ice felt the change as the deep liquid freshwater flowed outwards, forming the jet. That foundational ice was everywhere riddled with melt-holes and drainage channels the size of major rivers. Shifting forces caused by loss of water via the jet upset the foundation's delicate equilibrium of forces vs strength: the lower level shattered into monumental but mobile chunks, their movements lubricated by the lake water. As is true of most structures, diddling with the foundations is not a good idea. The deep fracturing propagated outwards in all directions from the jet's entrance, at over 100 kmph.

Then the ice above the re-arranged foundation began to settle vertically, the settling arriving at the surface quite quickly. If given the proper vantage point, one could have watched the surface effects of the deep collapse. As the underpinnings rearranged themselves, the surface layer of ice far above shook violently, rather like a cosmic-scaled hound might shake itself dry after a good swim. The surface layer to a depth of several hundred metres shattered and turned into a roiling, churning mass of ice of nearly unimaginable violence.

Some of the resulting spectacular near-surface breakup flushed almost instantly into the sea well north of Jayhawk… initially far out of sight from the ship, but the ice would quickly begin to show up on her radar.

The inhabited southwest coast received a torrent of water and ice well over 200 m thick, tens of miles long, and moving at over 400 kmph. Within a few seconds of the flood reaching the coast, no evidence remained to show man's recent presence. The GIC flood's excavations dwarfed those made by Lake Missoula, which today carry the Columbia River.

Structural collapse of the GIC's basal supporting structure had taken under two hours from the start of flow, and it took only a few more hours for the water and ice to get into the ocean… and a mere two percent of the GIC's total water and ice reached the sea.

EARTH'S OCEANS REACT

Within a day, Earth's MSL rose between fourteen and fifteen centimetres. That number is so small as to sound perfectly trivial. But in this context it is actually anything BUT trivial - after all, there are consequences to, and linkages occasioned by, changes in almost any environmental parameter. At least occasionally the changes themselves may be quite predictable, but the consequences of the changes are usually not so.

●●●

To the Boys' Club, Poseidon declared gleefully "Behold - the Mortals never understood the warnings, not even the third and most blatant. The humans did almost nothing useful! Truly amazing!"

Then it was, "Pay up, gentlemen!" Bets were collected, accompanied by resigned shrugs and sheepish grins from those who had backed the Mortals.

●●●

{T_0 + several hours}
COLLAPSE: GRAVITY WINS

On the inland side of the coastal mountains, the Cap's basement ice felt the change as the deep liquid freshwater flowed outwards, forming the jet. That foundational ice was everywhere riddled with melt-holes and drainage channels the size of major rivers. Shifting forces caused by loss of water via the jet upset the foundation's delicate equilibrium of forces vs strength: the lower level shattered into monumental but mobile chunks, their movements lubricated by the lake water. As is true of most structures, diddling with the foundations is not a good idea. The deep fracturing propagated outwards in all directions from the jet's entrance, at over 100 kmph.

Then the ice above the re-arranged foundation began to settle vertically, the settling arriving at the surface quite quickly. If given the proper vantage point, one could have watched the surface effects of the deep collapse. As the underpinnings rearranged themselves, the surface layer of ice far above shook violently, rather like a cosmic-scaled hound might shake itself dry after a good swim. The surface layer to a depth of several hundred metres shattered and turned into a roiling, churning mass of ice of nearly unimaginable violence.

Some of the resulting spectacular near-surface breakup flushed almost instantly into the sea well north of Jayhawk... initially far out of sight from the ship, but the ice would quickly begin to show up on her radar.

The inhabited southwest coast received a torrent of water and ice well over 200 m thick, tens of miles long, and moving at over 400 kmph. Within a few seconds of the flood reaching the coast, no evidence remained to show man's recent presence. The GIC flood's excavations dwarfed those made by Lake Missoula, which today carry the Columbia River.

Structural collapse of the GIC's basal supporting structure had taken under two hours from the start of flow, and it took only a few more hours for the water and ice to get into the ocean... and a mere two percent of the GIC's total water and ice reached the sea.

EARTH'S OCEANS REACT

Within a day, Earth's MSL rose between fourteen and fifteen centimetres. That number is so small as to sound perfectly trivial. But in this context it is actually anything BUT trivial - after all, there are consequences to, and linkages occasioned by, changes in almost any environmental parameter. At least occasionally the changes themselves may be quite predictable, but the consequences of the changes are usually not so.

●●●

To the Boys' Club, Poseidon declared gleefully "Behold - the Mortals never understood the warnings, not even the third and most blatant. The humans did almost nothing useful! Truly amazing!"

Then it was, "Pay up, gentlemen!" Bets were collected, accompanied by resigned shrugs and sheepish grins from those who had backed the Mortals.

●●●

2

GLACIOLOGY LAB, ASHORE

Kevin never did figure out what triggered his sudden, urgent need to drop by the glacier lab so early in the morning of $\{T_0\}$.

He'd slept fitfully – unusual for him – gotten up, dressed and headed towards the lab long before sunrise. Enroute on his bike, his own lights lit and reflective everything aglow in the light from passing vehicles, he decided to attribute that need to a generalized nervousness and excitement over this week's upcoming activities. Namely, the official, formal startup of their multi-year observational program to study the size, shape, composition and behavior of the whole Greenland Ice Cap, in 3-D. A hugely-difficult proposition.

The program of necessity was big and complex. Designing the research program, developing and testing the instruments and the complex programming that turned five hundred instrument packages into a coordinated system, getting adequate funding, building and deploying the packages, finding two months of icebreaker time – all that had taken most of a decade. The project was uncannily like satellite-based science, half a career invested before getting any data – and then only if the satellite didn't blow up or fail to reach orbit. Nobody had even thought about another underlying similarity between this project and satellite-based science… the fact that such a large, well-handled program of hundreds of instruments emplaced over millions of square kilometres could explode on takeoff.

At any rate, the scientific, political and other preliminaries were now behind them, and shortly the team would at long last push the "launch" button on the science. Their group's primary task was to answer this apparently simple question 'Interface: frozen solid, or liquid lake?'

As he dismounted in front of the lab Kevin realized that today - in fact right NOW! - Doc should be aboard one of the cutter's choppers, helping place the final few instruments.

The deployments had gone remarkably smoothly, overall – Murphy seemed to be on vacation for the moment (HOORAY! – plus 'knock on wood'!). Last year's 'western' instruments were up and operating normally – just two out of over two hundred packages had failed in over a year, a remarkable achievement. All the rest of the western-half packages were busy busy busy, just now finishing their first few months of shoveling massive amounts of data back to the lab via satellite link. The year just past, with only the first half of the instruments in place, had been dedicated not to actual science, but rather to getting the overall system ready for full-scale operation after this year's deployment of the second half of the instruments. The initial success of the western half had, of course, been a huge relief to everyone, from Congress and NSF's leaders, right down to the project's undergraduate lab-helpers.

Although observations were almost completely automated, the project had this year emplaced on the eastern side three four-person teams of graduate students, only accidentally all males – qualified not by gender but by having extensive cold-weather mountaineering experience. The live bodies had all been deployed via helicopter, and were on the island only temporarily – all would be retrieved (the Coast Guard, being paramilitary, liked the term 'extracted') about ten days hence. The teams were to collect physical samples, take detailed photographs, and do other ad-hoc work that was necessary, but either could not be planned ahead, or was too off-the-wall for automation. Teams B and C were operating inland, on the Cap itself, about forty kilometres from the eastern seaboard. Team A was not on the ice cap proper, but on top of the coastal mountains in the southeast sector of the island, setting up a remote, unmanned meteorological station – a system that required being bolted down, if it were to survive winter winds. Bolted to quite a number of anchors drilled by muscle-power into the mountaintop.

Membership in any ashore-Team was considered a signal (if obscure) honor - there had literally been a competition and lottery for positions. What budding geophysicist or glaciologist **wouldn't** want to go? Even if it required bashing twenty or so ten-inch-deep holes in rock, using a six pound hand-sledge and an iron-age star drill!? The A-Team's intangible rewards for their efforts included spectacular views to the west over the Cap, and to the empty sea eastward. The three Teams could communicate with Jayhawk via satellite cell-phone link, but had needed little comm to date - things were going smoothly hence they had nothing to report save success, easily and briefly chronicled.

Back ashore at the main lab, keeping track of just system operations, much less of results, was a major task… but Kevin had designed, and the group's electronics engineers had built, a lovely display, using a 72-inch flat-screen TV. When idling, the display sported a full-screen map of Greenland, overlain with satellite-measured topography of the ice cap).

Sprinkled across the entire map were almost 500 bright green dots – each indicated an instrument package that was up, operating, and on-link, sending data home in real-time. Any package having a self-diagnosed problem would display as a blinking green instead of a steady glow. At the moment, and very much "as usual", there were no blinkers. Once data began to be processed in earnest, all results could be overlaid on the basic distributional and topographic maps. In any case, even before starting 'official' observations the display provided a fine picture of large science, going well.

It had been a huge treat to watch newly-emplaced packages come on line, popping into existence on the map, each new dot helping to steadily shrink the remaining hole in the overall pattern. There were also three bright orange dots, indicating the A-, B-, and C-Teams.

Designing the pattern of package-placement had been a difficult, highly contentious process. A symmetrical, invariant rectangular grid of sensors had some serious advantages – as did less regular patterns designed to take account of existing knowledge... e.g., by bunching up instruments in areas of special interest. The resulting pattern, displayed on the TV, was a compromise: the distribution seemed rather chaotic, concealing from the human eyeball the underlying pattern regularities and the rationales for them.

PRELIMINARY ANALYSES

It would take at least a full year of 500-instrument operations to provide the necessary detailed data for anything like a conclusive analysis... however, earlier that very evening Dot had gotten the main analytic program to run (months ahead of plan!), and had tested it using preliminary data from some of the instruments in the first-year's deployment. She had sent an email about that accomplishment to the lab's party aboard Jayhawk, then gone home for a nap, which failed... too much adrenalin, she guessed. She returned to the lab – might as well get something useful done if she was going to be up all night.

{T_0 + 88 minutes}

FIRST ONSHORE INTIMATIONS OF TROUBLE

DotCalm (in reality Janis Symes, a name never used in her presence) let herself into the building again, via the keypad-lock, went upstairs one floor to the lab, entered it. Her opening the door roused Kevin from his reverie: as was his wont, he'd been sitting in full lotus on the oversize communal couch that faced the big display.

Dot had spent her entire graduate-student career immersed in this project, studying the GIC. Or more accurately, working on the technical preparations needed to get the data with which to study the GIC. At any rate, she was Kevin's and Doc's most senior -and very best- graduate student. She was also their top computer jock and famous for sangre-froid under fire. No matter what kind of uproar, or from whatever direction, she was unflappable. She was, consequently, nicknamed "DotCalm".

"Hey, Professor Kevin SIR! - You planning on sleeping here again? You're NUTS!"

Kevin stood, shook his head. "Not actually asleep. You know me, Dot... scientific monomaniac. I couldn't sleep, all keyed up. I decided to come over here so I could watch the final placements. It's been going very well... just like last year. Out there on Jayhawk they're nearly done, maybe 20 or so packages left to drop. Just half a day's work! I'll bet Doc is airborne right now, with the final few instruments. Might be symbolic, the Old Man Himself planting the last instrument in his magnum opus."

Kevin paused, yawned baboon-style, asked "Shall I make a couple of espressos?"

Dot nodded, said "Great idea!" Then, carefully nonchalant, she said "Oh by the way, I got the main program to run today, using some year-one data for a test. A mere 5 months ahead of sked – you can now appoint me resident heroine for a day and thank you very much!"

Kevin went to the espresso machine in the lab's tiny kitchen, turned it on: while it warmed, the scarlet beast hissed ferociously, like a baby dragon. Over its hissing came his response - "Congratulations! Way to go!"

She grinned to herself – she purely loved occasionally being able to steal a march on the rest of the lab. "It looks to me like the Interface Lake group is going to win... I think I can see a thick layer of water down at the surface of the rock. Needs lots more data and better analysis. Anyhow, Kevin, I sent an email about all that to the ship a couple of hours ago. No response yet." Then "I'll show you the output later tonight, maybe after the final instruments are in place."

Dot was happily contemplating the solid display of steady green dots, still waiting for the coffee, when a green light in the NE sector blinked several times, then disappeared.

"What the fuck?" she said out loud, staring. It had happened so quickly and in such a dense field of green dots, that she couldn't even be sure within a couple of inches of its now-dark position. The sudden disappearance certainly suggested total, instantaneous instrument failure. But to date that failure mode was unknown amongst the deployed packages. Even the two "failed" packages from last year's suite had only gone intermittent on a couple of sensors.

The espresso machine finished its task, hissed in a different note, less dragon-like. Kevin appeared, handed one cup to her, asked "I heard you just now. What the fuck is it with your 'What the fuck?' - over?" He grinned, "Explain yourself, Madam Heroine first class."

She pointed: "I just saw one unit conk out - it's gone now. Totally gone-gone, Kevin – not even blinking! That means the unit is dead. Un-fucking-believable! And it's one of this year's units… it's been deployed for under two weeks!"

It was, indeed, hard to spot any such tiny change – one dot - in the near-chaos of the 500-dot pattern. He eyed the splattered array of dots, said "I really don't see it. You sure about that?"

Dot nodded: "Yep. Right about here." She tapped the screen. "Blinked three or four times, then gone. Hasn't come back up, either. Maybe I'm over-interpreting, over-reacting. Let's see if it does the auto-reboot that we built in. Should take less than two minutes."

Two more, side by side, blinked near her finger: one blinked only once, the other five times, then both disappeared from the display.

"Jeezus Keerist!" said Kevin…"I guess neither of us is nuts!"

DotCalm said "Wait. Watch. Think! Probability number one is comm failure."

"Not likely to be a simple communications problem, Dot - all three were physically close together, but comm failures should be random, not clustered. Remember, the individual packages are independent entities. The units don't talk to one another, just to us."

Two more dots went blinkers, then disappeared entirely. "Fuck me! What the hell?" That was DotCalm, suddenly anything but unflappable.

Then came three more: the hole in the displayed pattern was now obvious, the size of a fist. In total befuddlement, they watched a steady stream of failures, always at the edges of the dead-hole.

"Shit oh dear, this is like watching a cancer grow!" said Kevin.

{T_0 + 95 minutes}

PATTERNS

DotCalm nodded, studied the pattern: they watched as several more packages gave up the ghost.

"Kevin, do you see what I see?"

He did.

They stared at one another, then back at the display. The newly-dark patch was growing fastest in the direction of an orange dot - towards Team C. It was DotCalm who finally spoke. "Kevin, there is only one obvious explanation for this. Something serious is happening to the ice cap. I'll bet it has to do with that damned Interface Lake --- what else could it be? Something is going on that's serious enough and violent enough to kill our instruments. Whatever it is, it's growing, and fast, towards Team C. Our guys are at least thirty miles from solid exposed rock, this thing, whatever it is, looks like it's moving, growing, at maybe twenty-five, thirty miles per hour. And it looks like it's gaining speed, too."

Kevin shook himself, stared at the screen as several more NE sector dots blinked out. He hemmed, finally managed to say "If what you say is correct, and it sure looks to be, then whatever is killing our packages is going to kill Team C as well. My guess is that you're right and that part of the cap has found a way to collapse, all of a sudden. The collapse is propagating. And it's not just moving in on Team C, it's also expanding to the south and west. Towards Team B."

The on-screen hole in the pattern was now football sized. From no hole to major, in under ten minutes. And not a single unit had so far come back on-line via auto-rebooting.

{T_0 + 98 minutes}

REALIZATION

Both of them knew every member of all three Teams – knew them well, socially, scholastically, professionally. Dismayed, suddenly gone uncharacteristically frantic, DotCalm turned to Kevin: "Who do we call? What do we do? There's no way we can talk to Teams B and C, or send a rescue chopper. This is all happening way too fast." She stared at him, ashen-faced: "Kevin, I think our friends are DEAD! That thing, whatever it is, is going to reach them in less than five minutes. All of our people are going to blink out just like the packages. It's GOT to be something geological that's going on – but how the hell can a geological process move so fast? It's like watching a volcano explode. Or maybe watching an avalanche, or a landslide - nothing happens until BLOOEY, and in ten seconds everyone dies!" Then, quietly, with sudden tears trickling, she muttered "They just don't know it yet." With a shudder and gasp, she turned from the screen, shaking, muttered "I can't watch this anymore." Kevin hugged her hard, trying to be comforting but not knowing quite what to do. He continued to watch the screen over her shoulder, himself now the mongoose in the cobra's gaze.

{T_0 + 105 minutes}

HALF GONE

Despite not wanting to watch, try as she might, Dot couldn't keep her focus elsewhere. Utterly mesmerized, the pair clung to one another, unable to speak, unable to tear their eyes from the screen, watching the cancer-like dead-space grow. The blank spot's rate of growth was obviously increasing, and it began sending out little tentacles of darkness ahead of the main blank area. Not a single unit cycled itself to come back on-air after going off – gone as if down a sewer.

{T_0 + 109 minutes}

THREE QUARTERS

Dot managed, finally, to speak. "Kevin… if we're right about what's going on, then it looks for sure like the whole icecap is collapsing. All that water and ice is going to be running downslope at I'll bet a couple of hundred miles per hour. Lots of it is going to go straight out over the western coastal plains where all the Greenlanders live. Nobody knows it's coming. The whole damned island is probably going to be scoured clean in the next twenty or thirty minutes. And you and I can't even issue a warning! What the hell could we do, who would we call? What could ANYONE do?!"

Kevin nodded, managed to croak "Yeah. Who? And what the bloody hell difference would it make if we COULD call someone? The Cap is turning into a million-cubic-mile meat grinder - downslope of something like this, nobody gets out alive. Nobody even gets FOUND afterwards! Shit."

{T_0 + 114 minutes: Team C}

TEAM C

The sudden, completely unexpected combination of impossibly huge non-stop noise, with its unknown source rapidly approaching, plus the shivering and violent shaking of the ice beneath their camp brought Team C together, stumbling as the footing shifted beneath them when the leading edge of the churning, cosmic-dog-shrugging, arrived. They clustered, watching as huge chunks of ice reared impossibly upwards, spun about, collided even in mid-air, and fell back to simply disappear from view. Waited while their leader tried to raise Jayhawk, their only contact with the outside world. He never got through. The four men of Team C died together, without any understanding of what had happened.

{T_0 + 115 minutes: Team B}

TEAM B

Team B's last few moments were identical to, and almost simultaneous with, Team C's. Team B, too, had been unable to contact anyone – if they had made contact, they could not have transmitted an intelligible message because of the 160 decibel roar of collapsing ice, churning chaotically as it began its fall towards the compacting foundation-ice three kilometers below.

{T_0 + 128 minutes}

FINALE

Out of a field of nearly five hundred instrument packages, there remained exactly three, steadily glowing, improbably clustered at the north-eastern-most edge of the former GIC, almost mocking the rest of the display. At a total loss to say anything emotional or sympathetic, Kevin stared at the triad, shrugged, snorted gently, and muttered "At least with all the ice gone, it should be a bit easier to map the goddamned basin!"

Kevin was wrong only about the "ALL".

3

JAYHAWK IMPERILED

Rebecca took yet another look at the high clouds – they were gone, suddenly cleared away by high-altitude Arctic winds. Satellite imagery would be possible now. About time! To do any serious planning or thinking about the freshwater phenomenon, they needed the very-large-scale context that satellites provided. Of course, this delightful atmospheric clarity could disappear in a mere blink – so she headed at once down to the radio room, where she dialed up one of the dedicated geosynchronous photo-satellites that could get good imagery of their local area. She typed in her personal user-ID, specified the desired viewing area, ordered up both visible-wavelength and thermal (a.k.a. infra-red) images, at a coarse spatial scale. Instantly she got an acknowledgement and time estimate – only a ten minute wait, amazingly fast, apparently nobody else in the queue.

Consumption of mugs of coffee is an integral part of the time-keeping system aboard any American ship. One half of a coffee passed before the files arrived. The infra-red pictures showed, extremely accurately, the temperature of the skin of the ocean. The puddle of freshwater was spectacular in the IR, being everywhere a couple of degrees warmer than the surrounding waters. In about two seconds of scanning by eyeball she found Jayhawk – her engines' exhaust threw off huge amounts of heat, so the ship was easy to locate, a big brilliant white hot-spot in the infra-red imagery.

Rebecca also found the apparent source region of the freshwater, a couple of hundred kilometres north of their position, at the very northernmost edge of the satellite's viewing area... the imagery didn't extend any farther poleward. "The Puddle" was distorted, greatly elongated from N to S and widening as it extended southward. It was 20 or 30 kilometres wide at Jayhawk's position, obviously moving south, riding atop the south-flowing East Greenland Current. But The Puddle's north end was stationary relative to the planet – not drifting at all with the current. Which meant that the north end of this huge flow of freshwater must be continually renewed, from something like a point-source. Most peculiar indeed!

The duty satellite also carried a multispectral imaging system: BOs tend to stare a lot at the green-wavelength images – green indicates chlorophyll (and by implication photosynthesis) in the near-surface water. Darker green = more chlorophyll. She brought up the chlorophyll image, shook her head in wonderment – the entire two hundred kilometres long puddle showed as ZERO in the green – just like the water she'd looked at earlier in the day with her football. All the surface salt-water surrounding The Puddle was loaded with chlorophyll, as it was supposed to be. And presumably the chlorophyll-laden water was also at normal salinity – salinity being an important variable not yet measureable from space.

She shifted from the multispectral back to simple optical photos, set the crosshairs on the brightest-white (= warmest) spot in the north end of The Puddle, zoomed in gradually. She quickly found herself staring at a fine, slightly oblique overhead image of a screwball feature that would shortly be nicknamed 'The Boil'. The feature was about 60 to 100m tall and 500-600m wide, seriously foreshortened in the picture.

For perhaps a minute that image meant nothing whatever to her, her brain scrabbling to latch onto the imagery and make it understandable. Then the penny dropped – she was looking at a monstrous vertical jet of freshwater! It had to be from the Greenland Ice Cap, somehow getting through the East-Greenland Mountains which contained the Cap on its east side. An entirely new physical/biological phenomenon – and SHE was going to be the first to report on it. She had a tiger by both its physical and biological tails. Whizzer!

{T_0 + seven hours}

ABOARD JAYHAWK

Jayhawk was at idle, just station-keeping until the second of her two HH65 helicopters (call-sign Meadowlark-2, the name being almost obligatory, given that the Western Meadowlark is the Kansas state bird) came back to roost: number one was already refueled, otherwise serviced, and stowed in its hanger.

Jayhawk's immediate goal for today had been to complete installing the GIC-covering network of several hundred instrument packages. Last year, Jayhawk's slightly-older sister ship 'Healy' had steamed slowly up Greenland's west coast, using her choppers to instrument the entire western half of the isle. Jayhawk was nearly done, now, with the eastern half – only five more flights with four packages each, and thereafter the ship could revert to more traditional, largely oceanographic pursuits – currents, floating ice, even the occasional biological investigation as led by Rebecca.

The leader of the glaciologists, Doc Adkins, was a member of one of science's rarest species, a rheological glaciologist. 'Glaciologist' is pretty much self-explanatory. "Rheology" is an oddity, the study of flowing materials – water, ice, plastics, slurries, gasses. Glaciers moving as slow-motion fluids were Professor Adkins' specialty. He was known to all Jayhawkers as "Doc", and at the ancient age of 68, he still enjoyed, and insisted upon helping in, the fieldwork. Including the grunt-work of simply planting his instrument packages.

ON JAYHAWK'S BRIDGE

Summertime nights at this latitude were ridiculously short, the depth of darkness hardly a matter of concern. But the northern midsummer dawn, such as it was, would be upon them shortly: the sky was already pretty close to its uniform grey-white backlit daytime self. Lieutenant Julie Jonson had the bridge – Captain's wake-up was the bridge-watch's task, due in about 45 minutes. This was Jonson's first cruise aboard a Healy-class breaker: although it was early in her career, she was already fully competent and qualified, widely experienced. Amongst other educational and training tidbits in her dossier was recent certification as an advanced radar-observer. A well-trained but newbie radar operator, she could be forgiven for being bored with that part of her duties – there were no other surface vessels of any size or persuasion within about 200 kilometres, well beyond the range of Jayhawk's radar. And for the past few days, horizontal visibility using the _Eyeball, Human, Mark-1 Mod-0_ hadn't been all that bad, really, probably about 15 to 20 kilometres at noon, from the bridge… making the radar superfluous, except for finding ice in the dark, or tracking choppers relatively nearby. Ice isn't that good a radar reflector, but then a King-Kong Class icebreaker like Jayhawk didn't need much protection-by-avoidance. In short, there was nothing out there, aloft or afloat, to return an interesting radar signal to break the monotony.

Coffee time, a quick walk outside onto the exposed wing of the bridge for a cold-air wake-up, then back to the radar. Next known, expected object of interest would be the returning #2 chopper. Although not specifically designed to look for airborne targets, Jayhawk's radar could see a model HH65 tens of miles away. But the bird still aloft wouldn't yet show on-screen, wasn't due back in range of the radar for another twenty minutes or more.

For a moment, Jonson watched the radar screen's classical ship-centered rotating sweep: she'd left the set in ten-nautical-mile range for the past couple of hours, normal ops when the ship was idling about, only stuff quite nearby was likely to be of any interest. If the small-craft were out doing their thing, or surface parties were on the ice, you could keep track of them with the 2, 5, and 10 mile ranges, but nothing of the sort was going on… just several oceanographic "over-the-side" instruments dangling on wires short enough to prevent them tangling, the scientificos making maximum use of available ship-time. Even when idled down, as now, the ship was a blooming buzzing hive of activity. On any research vessel – particularly something the size, rarity, and cost of Jayhawk, there was never any time without multiple scientific programs being worked, 25/8 not merely 24/7.

Coffee in hand she twisted the range knob – two clicks brought up the 50-mile range – might be able to see Meadowlark-2 by now, break the boredom. She stared: halfway out, range twenty-five nautical miles, there was the goddamnedest clutter she'd ever seen on a radarscope, extending completely across the screen. Not only that, but the screen showed solid clutter, like ground return on a shore-bound set, from the clutter's southern edge at range twenty-five all the way out to the display's max, now set at fifty.

The display wasn't static, either - different parts of the clutter blinked, winked, changed brightness abruptly, on several spatial scales. She understood at once – the clutter must be from an entire field of targets, apparently all in violent motion. Only big targets moving fast, and rapidly changing the orientation of their radar-reflective surfaces, could produce such a signal. But logic aside, that just didn't compute - the display simply had to be nonsense. Nonetheless, nonsense or not, the splattered, rapidly-changing green fluorescence almost shimmered, there was so much activity in it.

Jayhawk was holding station as planned, with Greenland to port and the ship headed due north into the screwball freshwater surface current they'd discovered earlier. The line of clutter, or interference, whatever the hell it was, butted right up against the return from the nearly-vertical coastal mountains. Lt Jonson flipped to 100-mile range: the sparkle-field extended farther to the north than the set could see the sea-surface.

"Shit. What the hell?" Jonson queried herself. She quickly reviewed her radar training. This had to be an equipment problem - she'd never seen or heard of any 'interference' or clutter such as this. Given some oddball behavior of the set, present situation qualifying, standard procedure was to switch to the identical, completely independent backup system. She did so. In ten seconds set B was up. The two sets displayed identical, rapidly-changing chaotic patterns of bright and dark.

In the middle of the 100-mile view of sparkles sat Meadowlark-2, her identification beacon beeping away perfectly – and she was closing Jayhawk at 180 knots, going flat out, balls to the wall. "Not fuel-efficient", Jonson thought… "I wonder what's got them buzzed up there?"

Puzzlement aside, she believed her instruments, and so avoided being a victim of Three-Mile-Island Syndrome. She drew the proper first conclusion – given two independent sets with identical displays, this was NOT a radar problem… a real-world phenomenon. But what?

Suddenly nervous atop her puzzlement, she flipped on the manual Doppler, to see if whatever was giving such a return was stationary or moving, and if moving, then towards or away from Jayhawk. She blinked in disbelief: positive Doppler –the clutter-target was approaching … and at over forty knots! Now, she could see the decrease in range by eye: on the radarscope, a fraction of a millimetre advance in Jayhawk's direction with every four-second sweep.

She looked forward over the bow into the brightening all-whiteness of sky and sea, could see nothing weird by eyeball. The horizon, from the bridge, was about sixteen kilometres away, so of course she couldn't see the target. Not yet.

Deeply ingrained in her through all her ship-handling training and experience was the fact that, for as long as mankind had gone to sea, absent a desire to ram somebody, "*Target bearing steady, range decreasing*" was NOT a good situation in which to find one's-self. And that was precisely Jayhawk's situation here and now. No matter what the hell was going on just north of them, whatever was giving this humongous signal would be atop them in a very VERY small number of minutes.

As Jonson reached for the bridge phone to wake the Captain, she had a shivery, knee-wobbling thought: closing speed with Jayhawk stationary was about forty knots. At emergency speed, Jayhawk could make just over seventeen. Maybe eighteen on a good day. "It", whatever it was, was busy coming to visit them, and would arrive in their lap quite soon, wanted or not.

And Jayhawk couldn't possibly outrun it. Not even close.

OhMyGod!

Jonson didn't hesitate after that realization: per her training, on her own initiative, even before the Captain came on-line, she sounded the raucous alarm for "general quarters" – meaning 'battle stations' for the Navy, and the equivalent in the Coast Guard – all non-essential activities stopped, all personnel to emergency stations and dressed in both life preservers and combat gear, every watertight door in the vessel slammed shut, dogged down and double-checked. Jonson had never before sounded GQ for real, but she (and every ship's officer) had had drilled into her the maxim "When in doubt, sound GQ!!" After all, a false alarm was a harmless embarrassment that could be called 'training' - but **missing** a real need for GQ could easily kill the ship and its entire company.

Jonson went further than merely setting GQ: she flipped the mike switch to PA, announced loudly to the entire ship "This is the bridge. Cease all over-the-side activities immediately! All outboard gear must be either gotten aboard or jettisoned in two minutes. DO IT! No exceptions. Engine room, all ahead emergency, con give me a 180 turn to starboard, right full rudder NOW, new course due south. MOVE IT, PEOPLE! GO! THIS IS NO DRILL!"

Captain Sheldon arrived on the bridge in under twenty seconds, jamming his shirt into his trousers, blinking away the final traces of sleep.

"What the hell's up, Lieutenant? General fucking Quarters? Are we under attack?"

Jonson motioned the Captain over to the radar, gave him a 30-second history and analysis of the situation, showed the second set for confirmation, explained her interpretation of the displays. She ended with "Sir, up here at this latitude there's nothing except ice that can produce such a return, especially where we're supposed to have only open ocean... and that much signal at this range means a TALL target. If it's ice, it's a 500-foot mountain on the move in our direction! And at forty knots, Sir!"

Sheldon sized things up quickly, looked at Jonson, said "Damn! You're right, of course. Sounding GQ was the perfect reaction, glad you didn't wait. Likewise getting us under way headed south. What closing speed on this fricaccatta target?"

"It's making forty knots, sir, and steady. Closing with us at forty minus our own speed. Closing should be about 20 knots with us at flank."

"Well hell! Whatever the bejeezus that is, it's huge and we can't outrun it. But I don't see anything else worth even trying! We can't go east fast enough to get clear of its path. God-damned thing can't be fog, must be solid, which means ice, which means it'll at least try to kill us when it catches us. And it's sure as hell big enough and fast enough to do so. Both. Catch and kill." Then, totally puzzled, "If that target is what it has to be, namely ICE, what in holy hell is pushing it along at that speed? If you'd asked me ten minutes ago I'd have sworn there was no way an icepack could go faster than the current in which it sits – this is CRAZY!"

He stared at his Lieutenant, feeling the ship's vibration change under acceleration. "If it's ice, and all of the sparkles on the scope are big chunks of moving ice, then absent a certified miracle, Jayhawk is 100% walking dead right now. Those individual pieces, individual echoes, must be over a mile across! Any other theories occur to you, Lieutenant Jonson?"

Jonson shook her head "no" – then said "But Sir… where in the world would that much ice come from, so suddenly?"

"Who can say? Maybe a big chunk of the icecap disintegrated somehow, and really FAST – earthquake maybe? We'd probably not notice a quake out here. No place else for that stuff to come from, and nothing else around here to give that radar return."

The Captain turned to his radioman: "Billy, you go squawk SOS loud and clear on every frequency you can muster, then get HQ on the horn, tell the bosses what's going on, what this looks like, and give them our location, course and speed. Tell them that until further notice, we will run at flank speed straight south away from the mystery object, which will overtake us in just about an hour if nothing changes. We cannot maneuver out of the target's path in the time available. If it is ice, we do NOT repeat NOT expect to survive any actual encounter." Captain Sheldon looked at his radioman, perhaps nineteen years old and gone pasty-pale. "Go do it son… we'll try to think of something."

Captain Sheldon then called all bridge personnel to the scope, explained. His analysis was brief, to the point. "Target is unknown, almost certainly ice, and in really, REALLY big chunks, moving like a banshee right at us. We can't make half its speed. If it slows down, we might outrun it, but it doesn't look like that's happening, Doppler is holding way positive, dropping just as much as the speed we put on. God-damned target isn't slowing a bit. If we can get 18 knots out of Jayhawk, that's a speed differential of 22 knots which means the thing covers our present twenty-five miles of separation in just about one hour. Ideas, ladies and gentlemen? Wild-and-crazy is probably a requirement. Try."

No instant ideas: into the pause, the bridge's 'ship to aircraft' radio lit off, going automatically into bridge-only PA mode. Everyone stopped thinking to listen.

"Jayhawk, this is Meadowlark-2. We are inbound, 50 miles, 180 knots, angels four." Then strict radio-chatter protocol went out the window. "Emergency emergency emergency, get on the horn NOW, guys!"

The Exec replied for the bridge - "Jayhawk here. Go ahead. We have our own emergency – state yours!"

"Jayhawk, our emergency is NOT repeat NOT an aircraft problem. We are flying above the goddamnedest thing we've ever seen. It's a huge floating ice-field, completely broken up into chunks, some of them easily a mile long and wide, and it's churning, looks like a mountain river in flood. Pieces of ice that big are coming shooting up through the pack and going way up in the air, maybe 800, a thousand metres exposed, no idea how much still underwater. The front looks like the wildest-possible head-on view of the world's largest snow avalanche. Just scaled up by a million. The ice field is mostly close-packed VERY heavy ice, the field is at least 50 miles wide, and extends to the north as far as we can see, at least 30 miles. It's coming your way at what we guess is 30-plus knots, give or take a few. I doubt seriously that Jayhawk can survive any encounter with this thing. Over."

"Meadowlark-2, copy. Seems there is just one BIG emergency today. We are aware of your ice-phenom and we are making best speed due south, but even so, closing speed is 22 knots, it will catch us in about one hour. Say your status and fuel, over".

"Aircraft 100% nominal, all packages deployed on schedule. Fuel 1150 pounds, about 70 minutes of flight available. Shall we land, refuel?'

The Exec, himself a chopper pilot, shook his head, said aloud to the bridge, "Guys and dolls, even with a full tank Meadowlark-2 couldn't get to within a thousand miles of anyplace useful." Then, "What good could it do to refuel #2, anyhow? Fill her up and launch her, then her only landing spot gets pooped and sunk as they watch from the grandstand. Bloody fucking hell!"

The Captain took the mike, keyed it: "Meadowlark-2, we have about sixty minutes, max. You stay aloft for the moment. Tell us about anything useful you can see. For sure Jayhawk won't take kindly to being sat upon by a square mile of airborne ice. That's beyond her design specs and way outside her performance envelope. We are polling ourselves for bright ideas. You have a different view from up there – got any wild-and-crazies for us?" He snorted: "Beers all around if you get one that works."

"Jayhawk, Meadowlark-2. Give us a couple of minutes to chat amongst ourselves. For about 30 minutes now, enroute home, we've been watching this thing move and churn and travel south. It does some amazing things, almost like it's alive. Back on the horn in no more than five minutes. Meadowlark-2 out."

ABOARD MEADOWLARK-2, AIRBORNE

LtJG Marcia Jameson, pilot, flipped up her face-plate, did a quick eyeball-to-eyeball inventory of her co-pilot plus four crewmen. "You all heard the Captain. That crap down there is going to kill Jayhawk, therefore us also, in less than one hour. They can't outrun it, and it's not slowing much if at all. From way up here, we have the best view of the overall situation. We have more fuel than we will ever need, if we lose Jayhawk. I'm going to engage autopilot, and we are going to confab. Ideas?"

Chatter mixed with silences for almost a full five minutes.

Because Meadowlark-2 had been busy deploying science-gear, the fourth crewman was not the usual "Coast Guard emergency swimmer-diver" but instead a rather skinny, easily-made-airsick and much older scientifico – namely "Doc". He stared out the window at the mountains that edged the big island: mountains a mile or more high, insanely steep, only about six or eight miles away at the moment. "Doc? Ideas?" queried Marcia. "Now would be a fine time to display your genius!" It was a friendly rib, allowable only due to long periods of working closely together.

Doc grunted, said "Maybe. Something is tickling the inside of my skull. Hand me the big binoculars. I need a close look at the entrances of a couple of our nearby fjords."

The copilot was using the binoculars, announced "I have a visual on Jayhawk… never seen her leave such a wake. Mega diesel-exhaust, too. Making knots like a bat outta hell, she is."

He handed the binox to Doc, who hummed to himself, brought them up to his face, and peered westward through the window. A minute passed. Then, suddenly, he let out a whoop. "Got it!" he said, exultantly – immediately tempered with "Well, maybe…"

The pilots looked at one another, then focused on Doc. Waving towards the shore he practically shouted: "Think! There are lots and lots of these fjords, we've been ogling them in flyby mode this whole deployment. Marcia, take us to the coast a few miles ahead of Jayhawk, NOW. We need to find a very narrow fjord that extends inland a couple of miles at least. Not too narrow, though, its entrance has to be plenty wide enough for Jayhawk. Most of these around here should qualify, they just look narrower than they really are."

Marcia disengaged autopilot, swung towards the coastline before asking "OK, Doc – WTF,O?"

"Wild and crazy! Remember, guys, "rheology" is me, the study of fluid flow. I was looking at the streamlines of the current as it flows past the fjord openings, the flow patterns are nicely outlined by the brash ice moving with the current. The current is southerly along here, and ***no surface ice is making a turn and getting into those fjords' entrances*** over there – the floes and bergs just scoot on past the openings. If there's a narrow fjord close enough, Jayhawk can duck into it… it has to have a relatively narrow mouth, but adequate for the ship. This mess of large-chunk shattered ice can't possibly turn a corner and go into such a fjord--- the icepack will just sail right on past a narrow slit entrance. Probably snag a bit on the mouth, then plug it up a thousand feet high, but I don't think the ice can get inside the fjord. Most of the pieces are far too big, and headed south anyhow. It won't push in beyond the entrance."

Doc paused, thought, then continued: "For yucks, let's say we have fifty minutes, equals five-sixths of an hour. Jayhawk is making about 18 knots, five-sixths of an hour at eighteen is fifteen nautical miles, which is the maximum total distance Jayhawk can steam before getting caught, but she's going to have to burn some time making the distance to the coast… maybe she has forty or fifty minutes from NOW!? Call Jayhawk and I'll explain… you two up front have to find us a tight garage that Jayhawk can reach. Gotta leave a bit of time, say five, for maneuvering and all that. We have a small search radius, strict criteria, and time crunch, assume a maximum of about fifty minutes available. Start looking, I'll deal with Jayhawk!" Then, quietly, "If we have the time and can make this work, it's going to be one VERY damn close exercise. We get one shot only. If and only if we're luckier than we deserve to be."

It took Doc only about twenty seconds to explain the scenario to Jayhawk's bridge crew: Captain Sheldon issued the highest verbal praise - "Bravo Zulu, Meadowlark-2." BZ is naval shorthand for "Fine job, well done." Then, "DO IT, Meadowlark. And keep us apprised, give us a steer if you even have a FEELING that one of the fjords within range might do. We've drawn a blank on ideas down here."

There were plenty of candidate fjords. Meadowlark-2 closed to three miles from the coast and eight miles ahead of Jayhawk.

"THERE!" said Marcia in a carefully-controlled voice, pointing. Doc peeled himself out of his safety harness, stared forward through the binoculars. "Good. It's hell for narrow, looks like twice Jayhawk's beam, maybe three times. That's great… assuming the Captain can drive well at flank speed. It also looks like a good, long, open stretch inland, and the body of the fjord is quite a bit wider than the entrance. Maybe we'll even get some maneuvering room? Or at least have space for braking!?"

Doc surveyed the relative locations of Jayhawk – in the middle distance – and the fjord's opening. "Not just close, but tight, too. The entrance has a big offshore chunk of rock that's going to make getting turned and then lined up with the fjord itself a pure bitch – don't want to bounce Jayhawk off the walls too hard! After all, we're going to need somewhere to land, pretty soon. Preferably something both afloat and not too seriously bent."

Doc turned to Marcia, said "Can we give them a mark, hover so they can turn around us, around our location, to make the entrance? Maybe we can chuck one of our floatable smoke grenades out the door from a hover, give him a better mark? He'll need any help we can give him, because that damn rock on the right is going to completely hide the entrance until they are almost into it. They'll have to turn before they can even SEE the entrance. No time for much course correction, since they sure as hell aren't going to slow down on final approach! This exercise will be done at full speed until full stop."

Pilot Marcia shrugged, said "I'm a certified Airdale, and I don't do ships… Doc, you need to tell me where to hover this bird – I can and will put her wherever's right for being a marker. Affirmative on the smoke, good idea. Get one out and ready."

Doc raised one eyebrow, looked back and forth between the pilots who were still completely puzzled and in the dark. He said to them, with a huge grin, "I used to do ocean-racing, in big, ocean-sailing yachts. Lots of it. This exercise is just like coming up to, and turning about, a race-course marker buoy. Except of course, the level of difficulty and the stakes are both a bit higher." He keyed his mike: "Captain, this is Doc. Make your course 205 true NOW. We have a candidate fjord, the only one available within reach, and it looks good. A killer of an approach, though. You won't be able to see the entrance until you come about around a big rock partially guarding the starboard side of the opening. Entrance's width looks to be about three times Jayhawk's beam. What is Jayhawk's turning radius at your current speed, with maximum rudder?"

There was a long silence from Jayhawk, then the Captain said: "That figure is half of what is called the ship's "tactical diameter", Doc. For Jayhawk at eighteen knots it's right around 350 metres, with full rudder and the ship heeled over about 30 degrees. It is one truly spectacular maneuver and nobody gets to practice it very much. Doc, you sound like an old-fashioned sailcloth-and-teak sailor... are you proposing to have Jayhawk whip a flank-speed maximum deflection blind turn around a mark and into this fjord of yours?"

"Yessireebob, that is precisely the idea. Hope you don't mind driving fast, Captain. Things are going to be a bit close. If Jayhawk slows down, we all lose! In your last two or three minutes of approach you should set up on Meadowlark-2 for the turn – Lieutenant Marcia is presently moving us to hover at the pivot point of the turn you need to make. As in yachting, Captain - Meadowlark will be your mark, pass the mark close aboard and begin your turn when the mark is abeam midship. When you call for it we will drop a floating red smoke grenade to give you a better mark on the surface. Once you come about and make the entrance, you have a couple of miles of rapidly-narrowing fjord to slow down in. Captain, when you straighten out at the end of the turn, you'll have maybe a dozen seconds to do course correction: at the end of the turn you should be steady on a heading of about 285 true, which should put the center of the entrance, and the long axis of the fjord, dead ahead. The entry looks completely clean, and we can see nothing subsurface along the fjord's centerline. Can you stop in under two miles?"

"Affirmative on the stop – we won't still be at 18 knots after that turn, but from eighteen we can back down to zero in about half a mile – stopping's the least of our worries, Doc!"

The Exec had been studying the rapidly-approaching coastline through binoculars; he reported aloud: "I have Meadowlark-2 in sight, altitude about 50 metres and holding. I see the rock. Cannot see the entrance at all, down at sea level. The tops of the mountains make the fjord look pretty damn narrow!" He handed the binoculars to the Captain, who studied the situation, then suggested to Lt Jonson a small course adjustment.

Only a few of the bridge crew had yet turned around to see their pursuer. Those who did, stared into a charging, foaming hill of enormous ice-chunks, a hill in continual motion, and which seemed to rise forever upwards, hundreds of blocks of ice, each one far, far bigger than Jayhawk, all of them being rolled, tumbled, submerged and then surfacing like gigantic submarine-launched missiles. The individual chunks of ice were huge, and the churning appeared to the eye to be going remarkably fast, despite the forward speed of the mass. The turbulent leading edge of the rolling disaster was by now only a few thousand metres behind them, a very small number of minutes remaining –single digits- until Jayhawk would be overrun, pooped, and sunk in perhaps ten seconds flat. The outrageous thundering roar from the pack was beginning to make voice communication difficult. And the pack was now gaining quite visibly, as Jayhawk ran down her last couple of pre-turn kilometres. Radar imagery plus the sound had made things behind them quite real enough for those who chose not to look.

Another long pause from Jayhawk as she steadied on the Captain's final course, set to just barely leave the view-blocking guardian rock to starboard. Then the Captain spoke to Meadowlark-2, patched into both the bridge's and the ship-wide PAs: "Hey, Doc… glad you're up there to help. This could almost be fun, you know… most modern sailors never get a chance to wring out a ship at maximum anything. I learned to con ships in the US Navy, in tin-cans and light cruisers. But that was about thirty five years ago. I've had lots of practice at tight, fast, high-speed maneuvers, in fact mostly over 30 knots, not quite Jayhawk's specs. Lots of practice but a long time ago. Guess I'll con the ship myself for this run, and if it doesn't work the Coast Guard can dock my retirement pay for the damages. We have in sight both the top of your fjord, and the guardian rock. Likewise eyes on Meadowlark-2. I'm being urged to look astern. Apparently the mess is pretty damned close and spectacular. I'm not looking. Hope someone is video-recording this. Folks, I sure would like to put this maneuver on YouTube next week!"

Then, formally, "Lieutenant, I'll take the bridge now – my con. And I promise not to leave too much paint on the damn rock! Now – begin radio silence except for Doc and me. Three minutes to our turn – the leading edge of the pack is about seven or eight minutes behind us … at most!"

The Captain ran through his mental picture and calculations again, making sure he had allowed for the several seconds it would take to swing the ship's massive rudder to full deflection: likewise that he hadn't forgotten that due to the freshwater current, Jayhawk, at eighteen knots relative to the water, was actually making better than twenty knots over the ground. A complex undertaking, this turn, and to be done entirely by gut instinct based on experience – no calculators involved. He eyed the scenario, studied Meadowlark-2 for a moment, eased the bow slightly to port, grunted, shrugged, and announced "Two minutes, ladies and gentlemen."

"One minute."

The Exec keyed the ship-wide PA: "ALL HANDS! Brace! Brace! Brace! This is not advice, this is an order! Maximum turn to starboard coming in sixty seconds. Expect about thirty, maybe even forty degrees of heel."

Standing wide-legged and braced at the control console, the Captain studied the passing rock, considered the course Jayhawk needed to be making as she exited the turn, looked at his hovering guidepost, took a deep breath, and went on-air: "Meadowlark-2, make smoke on my mark – four, three, two, one, MARK!"

A dot, trailing a thick vertical trail of dark-red smoke, magically appeared ever so briefly beneath the chopper, floated when it hit the water, spewing a dense red cloud. A few seconds later the Captain eyeballed the cloud as it came down Jayhawk's starboard side, muttered "Good show!" and without touching the throttle gave Jayhawk full deflection right rudder. He was good to his word, passing the guardian rock down the starboard side so closely that the water between ship and stone geysered up, reaching well above the main deck.

As the prow began its swing and the ship heeled over more and more sharply, the Captain was humming to himself and harkening back to much earlier in his career, to a barely-avoided high-speed collision with a major stationary -and utterly immovable- object. "Always better," he was thinking, "…to steer by what you can clearly see and judge, rather than worrying about the invisible other side of the opening, or about trying to be dead-center in whatever passage exists."

"And after all," he thought, "… today's hidden entryway is supposed to be plenty wide – at least thrice Jayhawk's beam, wasn't it? That datum came from an apparently-reliable semi-nautical source, namely Doc. Should be a piece of cake."

The Healy-class had been designed for maneuverability from the get-go. Jayhawk heeled over like a Hobie-Cat, her stern hiked far to port, the ship plowing sideways, her wake churning and definitely not going the normal direction. The vertical coastline was only a couple of hundred meters away and swinging past the jackstaff far too quickly to focus on. As Jayhawk cranked around, the left edge of the fjord-mouth came into view: the Captain followed his own advice, eased off on the rudder and aimed for what he could see, laid the left-hand face of the entrance ten feet off Jayhawk's port side at over twelve knots, straightened out perfectly as the right-hand face came into view. Jayhawk wasn't centered on the opening, but was running at precisely a right-angle to the entrance, headed inland.

"Jeezus Fucking Christ Almighty" breathed the Exec as the ship, having scrubbed off almost half her speed in the turn, settled with her centerline precisely atop that of the fjord. She was already regaining speed, getting away from the entrance and whatever form of mayhem might descend there.

Two minutes into the sanctuary, and it was obvious that the ship was running out of fjord – the sides were closing in fast. Captain Sheldon called for full reverse thrust. The big ship settled deeply by the stern as the power came on. One minute of backing down and the ship's turbulent, frothing prop-wash -now going forward- had reached midship - half the ship's length. She was officially dead in the water at that moment. The Captain called for zero turns, faced his Exec, shrugged, and said "Sort of like riding a bike – hard to forget. Nice of the Navy to let me practice for this!"

With the broken tension came a spontaneous round of applause, immediately squelched by the Captain - with a non-regulation grin on his face. Then, "XO, somehow I doubt that it'll be possible to anchor here, between two vertical cliffs and with I'll bet half a mile of water under us. Lieutenant Jonson, you have the bridge and con again; your job now is to get this vessel turned 180 to face outwards. Please do not bend either end of our pretty ship in that exercise. Like me today, you may someday be glad you had the practice, who can say? Then, set up the big fenders to protect the paint when we drift into one of these walls, get our station-keeping crew in gear. And oh, yes – let's land Meadowlark-2 – bet they're running a bit short on gas, getting anxious. At least they got the best view of things, that's for sure."

The Captain watched Meadowlark-2 begin her approach, laughed happily, said "I'm in deep trouble, XO – unless you have a cold six-pack handy? Remember my promise!" (Jayhawk, like all USCG vessels, is of course alcohol-free. It says so in bold print in the owner's manual – somewhere.)

The Exec laughed – he was the ship's acknowledged beer expert:, an indispensable advisor for all ship's social functions. "Can do, Captain – in the little reefer in my quarters as we speak. Maybe Billy could go get six and meet the whole aircrew when they deplane? Seems appropriate."

After the flurries of activity retrieving and servicing Meadowlark-2, and getting Jayhawk properly protected with fenders, things quickly calmed to a near-normal pace.

"Next, Number One, inasmuch as we're clearly going to be here for a while, no matter what, it's already time to begin planning for an extended high-latitude vacation. Let's announce an all-officers pow-wow in the officers' mess in forty-five minutes, and include Doc and Rebecca – they can handle the scientific party. Then we'll have an all-hands meeting in two hours in the main mess. From down here in this chimney, getting radio contact with anyone is going to be difficult. Since we ducked in here we've had no satcom, and nothing from the Teams, either. Ideas are welcome."

A mere 1400m from Jayhawk, the dense, churning icepack that had been a significant fraction of the Greenland Ice Cap growled and thundered and shrieked as it sailed harmlessly southeastward, past the mouth of what was already being referred to as "Jayhawk's Fjord".

While Lt Jonson was gingerly maneuvering, rotating her 16000-ton charge, the ship's communications folks were trying unsuccessfully to establish any sort of radio contact with the outer world. As expected, that proved to be impossible, inasmuch as the ship was basically at the bottom of a very narrow bottle with vertical sides well over 2000m high, a natural and very effective channelizer of radio energy – unfortunately pointed the wrong direction. Most of the ship's communications relied on line-of-sight contact with geostationary satellites, and there did not happen to be such a creature in sight, as one looked up the bottle's neck.

As the ship eased into its new orientation, the Captain complimented Jonson on a job well done, then said to the entire bridge watch, "You know, troops, now that we're properly parked, with the pointy end the right way, next things are (a) survival and (b) communications. We have no idea whatever how long we might have to hold out in here, no clue whether and when we might get out on our own – and due to all that ice outside, we don't know when HQ can mount any sort of relief or rescue mission. So we're all going to have to start thinking maximum conservation of everything – especially food and fuel. Let the ship's internal temperature drop down to say 2 or 3 degrees C – everyone can wear their fancy thermal clothes. We all of us get to recap Ernest Shackleton's little Antarctic junket." He snorted gently: "Well… sort of!"

"But the number-one task for the moment is making contact with HQ, and with one of the photo satellites – the big boys back home are going to wonder what's happened to us, we need to let them know, and we need pictures of whatever the hell is going on with the ice cap. Also likewise pictures of the ice right outside our door, here." He shrugged, said "The only thing I can imagine is that at least a big chunk of the cap was unstable and has collapsed. Hell, folks, the whole damn cap might have come apart – photos may help us decide what, if anything, we can do about our situation." Sheldon pointed to the several hundred metres of ice blocking the entryway to Jayhawk's Fjord, shrugged again: "Yeah, sure, we're an icebreaker – but unless Mother Nature moves that plug, we're stuck. And even if we weren't locked up tight, we'd need to have good satellite photos in order to survive out there. That is definitely NOT a ship-friendly environment. Personally, I'm all for trying to get comm going ASAP by whatever means our electronics group can dream up."

Then, quietly, "Getting comm includes trying for contact with Teams A, B, and C." He looked around the room. Lt Jonson spoke up and voiced what everyone was thinking: "Sir, my own feeling is that anyone who was out on the cap, not on solid bare rock, is almost certain to be dead. My guess is that Teams B and C are gone – they were fifty kilometres inland, nothing but ice for miles in all directions. Team A might have had a chance, since they were atop the mountains." She paused: "But of course we have to try, whatever the odds."

By the time the all-hands meeting convened, the electronics folks had dreamed up an approach to re-establishing comm with the rest of the universe. Their leader, scientific e-tech James (a civilian), explained the idea: "Not hard to do electronically. We can use a Meadowlark to take one of our portable radio relay units up to the summit, use the unit to link to a photo satellite, the voice and data channels, and the frequencies used by the Teams. The portables are good units, can handle all the frequencies we need and do it with good data-rates. They use almost no power, batteries should last easily six, maybe up to ten weeks even at the temperatures on top of these mountains. We know how to staple a unit to the rock, no problem – it's only about the size of a basketball – we can shoot threaded 3/8-inch molybolts into the rock with a .38 caliber Ramset bolt-gun, use them as anchor-points. We just need one worker to install it: twenty minutes of on-site work-time should be plenty. Because I've been there and done that already, I volunteer."

The Captain looked pleased, scanned the group, called for comments and got only murmurs of "Good idea". He looked at James and asked "Can you really install it single-handed? Be serious."

James replied at once: "I've done the whole installation solo, several times. Permission to do it, Captain?"

The Captain nodded to James, looked at Marcia: "Okay. Lieutenant, you fly this mission. Be sure you are fueled to the gills. For conservation, this will be a multifunction operation. Take James and his relay unit up topside, find a good site, test the gizmo before you actually install it. When it's been installed, as soon as it comes up, we'll do two things – from here, we'll contact HQ briefly just to let them know we're ok. Then we'll try to contact A, B, and C Teams. If we make contact, we'll vector you to to their locations – you'll have plenty of fuel to make one round trip. If anyone is alive, you'll pick them up, closest first, warm bodies only – no gear salvage. In any case, on your return make a good detailed recon of our fjord's plug and of the ice and ocean conditions out to say 25 miles. James, you carry a video recorder and get detailed pictures of the ice and especially the plug, so we can see what we're dealing with. Two pilots plus James leaves enough payload for four passengers per trip - if we're that lucky."

It took under an hour to check out and re-battery the relay unit, gather the materials and tools for installing it on the rock, and draw up detailed plans for the mission. Takeoff and climb-out were routine: a quick pass over the plug made it very clear indeed that it would be Mother Nature –not man- who removed that obstruction, if-and-when.

Then, as the chopper headed upwards towards the mountain crest, the floating new ice-field hove into view. Marcia's report was short and succinct: "Jayhawk, Meadowlark-2. The ice field outside the fjord is totally chaotic; for as far as we can see, this ocean is completely covered with ice floes and bergs, to an unknown depth – basically we see no open water whatever. It does look like the whole mass is moving, it's kind of restless, a bit like kittens under a sheet, but not the high-speed turbulence we saw when we were running for cover. I think the flow is already calming down. Big chunks still prevail – by 'BIG' I mean of unknown thickness and up to over a kilometre square. No room whatever in the pack for Jayhawk, even if she could get out of the fjord. If the ice is all drifting south the way it seems, then it'll wash out of our way, eventually. The North Atlantic is in for a big surprise! We're heading to the nearest usable mountaintop NOW. Meadowlark-2 out."

Minutes later, as Meadowlark-2 cleared the mountain-top, her crew got their first view of the results of the collapse. Marcia had been flying twice-daily missions over the island for weeks now, and had fine, detailed memories of the Cap as of even this morning. But the Cap had changed, utterly beyond belief – and in only a very few hours. Her memories of the Cap were no longer of any relevance: she would have to start over, from scratch.

Yesterday, the surface of the Cap had been merely heavily fissured to the point of impassability, but with large expanses of relatively smooth surface –often kilometers in breadth- between adjacent fissures. The Cap's surface had butted smoothly up against the west side of the eastern mountain range now immediately below Meadowlark-2 ... the range containing Jayhawk's Fjord. And yesterday, the ice had ended at the range's north-south ridge-line, leaving only the skimpiest traces of exposed rock at the very crest.

No more.

There was now at least several hundred meters of naked, nearly-vertical rock between ridgeline and the completely restructured ice-cap surface - a surface that seemed to have simply dropped away from the rock, as if it had slumped on an enormous scale. The newly-exposed west-facing bare rock wall extended as far north and south as Marcia could see.

She stared, and it took perhaps ten seconds before the new topography and anatomy of the scene finally snapped into focus. In addition to the ice-field having slumped several hundred metres vertically, the surface texture of the cap had also changed dramatically… transformed from a relatively smooth plain into an utterly chaotic field of huge ice jack-straws and uplifted, shattered blocks. Individual blocks were of all shapes, all sizes from suburban garage to miles across, all orientations from gently tilted to vertical – some blocks had even been flipped like pancakes, coming to rest upside down.

She shook off her shock and curiosity – she had a chopper to fly and a mission to complete. But first, she had to report all this restructuring of the Cap to Jayhawk.

Marcia described the sight -as best she could- to Jayhawk's bridge. When asked by Captain Sheldon for details and impressions, Marcia described the Cap's new configuration, knowing that words were largely inadequate… none of her listeners could possibly visualize even approximately correctly the scene she was describing. Stuck for words, she promised to return with plenty of video. She suggested that someone, maybe Rebecca, should get online ASAP and download before-and-after satellite pictures – that sort of overview was obviously going to be needed. She let herself ask questions out loud, as asides: "Where did all that ice GO?" - an awful lot of ice had gone somewhere and done so unbelievably FAST. How much ice had gone missing? How far did it get? Was the remainder stable or was this just the first stage of a grander phenomenon? If it was "merely a big slump" then how should/would it behave? Importantly, had the moving ice and water gotten all the way to the west coast, where all 40,000 Greenlanders lived? If so, the real question would almost certainly be 'Did anyone survive?' rather than 'Was anyone hurt?'

She paused: nobody aboard Jayhawk said anything. Finally she started up again: "Jayhawk, I'm afraid we're going to have to tell Doc some really bad news. To me it looks highly likely that the entire top layer of the Ice Cap has collapsed and seems to have run off downslope, with his whole project aboard. The original surface is GONE, and the new one is jumbled ice who knows how thick, and all mixed up---the new surface is the result of about a thousand feet of vertical fall. I cannot imagine any of the instruments surviving this…"

Then, quietly, "Also, I'm sorry to say it but there's simply no way anyone from B and C could have survived."

Taking advantage of Meadowlark's altitude, Marcia also tried, pro-forma, to raise all the teams on the teams' own frequency, but got no response, not even from mountain-top Team A. Assuming that the whole surface of the Cap had been affected equally, the conclusion was inescapable; everyone actually on the icecap itself - all the members of Teams B and C - had surely been killed. Neither Marcia nor James had any doubt that Team A was also gone, making the casualty count twelve scientists MIA/KIA.

Marcia shook herself. She and James and Meadowlark did have their immediate mission to handle, and at the moment, they had to find a place for the relay unit.

JAYHAWK'S BRIDGE

Captain Sheldon and Doc stood alone, together, off in the port wing. There was no color in Doc's face, and he was silently quivering. Sheldon draped an arm around Doc's shoulders, said nothing. Minutes passed. Finally Doc shook himself, looked at Sheldon, and said bitterly, "Long time ago, I was a Captain in the Marines. Company commander, grunts. Infantry. Two tours of pretty serious combat, Iran and elsewhere. Had to write a lot of letters to spouses and families, telling them what the fuck happened to their guy. Thought for sure that that part of my life was over and done with. Guess not.

Fuck." He paused, Sheldon said nothing, just waited. Doc finally continued: "Those twelve students are ...no, I guess that's WERE, the cream of the crop. A whole next generation, gone. They didn't sign up for THAT kind of duty, being wiped out without warning or purpose. If you're on active duty in the Corps, sure, it's an inherent part of the job, of the environment. But NOT in goddamned graduate school while studying ICE!"

He looked at the Captain, who said "Been there and done that letter-writing thing myself. I'll help best I can. Let's start with a good stiff drink in my cabin, as soon as things calm down- is that okay with you?"

It was as okay as one could make it. Which wasn't very.

THE MODIFIED GIC

For as far as the eye could see from Meadowlark, a vertical kilometres of ice and water had vanished like a conjurer's trick on a continental scale. Impossibly gone, leaving sheer, newly-naked, almost-vertical rock surfaces facing, plummeting hundreds of metres. The more-nearly-horizontal lower rock surfaces were littered thinly with fragments of the Cap, randomly-oriented ice-chunks up to the size of skyscrapers. Some, perched on slopes, looked like high-divers peering over the edge of their platform, waiting to be jiggled into motion, perhaps for a nonstop bobsled ride to the western coast. How fast such a block of ice would be going when it finally hit water, god only knew!

MEADOWLARK'S MISSION

On the very summit above Jayhawk there was a reasonably flat space, dining-table sized, with a clear line-of-sight to Jayhawk, more than two miles below them. The space was neither flat enough nor large enough to let the chopper actually set down, but seemed otherwise adequate for their purposes.

Marcia asked "Will this do?" and got a vigorous nod from James. He was experienced at working in cold from a chopper. He sat himself and his net of gear by the door, ready to exit on her command: as usual, she was smooth – and it helped greatly that there was precious little wind. Per the plan, she pulled away a hundred metres to hover while he surveyed the little site on his hands and knees, limbered up the Ramset. It took only about five minutes to shoot in the ten bolts, then five more minutes to lay out the cable tie-downs, get the gear clipped to the mountain, turned on. Moments later, James and his tools were back aboard Meadowlark.

Marcia called home: "OKAY, Jayhawk, we're done up here. The relay unit is installed and turned on, looks ops normal – you guys should check the frequencies, see what you can get. We'll hang here until you tell us how things look from your end. Over."

"Meadowlark, Jayhawk. We're ahead of you. We already have contact with the satellite net via your unit: give us a few minutes and we'll let you know for sure, but right now all seems to be working just fine. Meanwhile, be sure to get plenty of video. Hold steady where you are."

The bridge radioman announced, "All set to talk on HQ's frequency, Captain."

Sheldon took the mike, cleared his throat, and spoke. "CGHQ this is Captain Sheldon of Jayhawk. We are working a jury-rigged radio-link. Do you copy? Over."

A long and incredibly pregnant pause, then a cackle and a prolonged raucous "WHOOP! WHOOEEE!" – later traced to a Texan exported to CGHQ as a radioman. "HOT DAMN, Jayhawk! Jeeeezus – By Jehosephat, Sir, we thought you all had to be fucking DEAD! Please excuse my French, Sir, and say your position and condition so I can tell the bosses – without some such info they'll never believe me, please and thank you very much indeed, Captain! HQ over."

"Roger that, HQ, you can go tell the upstairs-folks that we outran the ice by about two minutes, and are safely parked in a very narrow fjord which protected us from the ice, but deep ice has completely locked us in. The fjord's entrance is plugged with ice about 200 m high. We'll send coordinates and you can get a satellite shot of our new garage! No damage, no casualties – except for three four-person shore parties which we have been unable to contact. All 12 souls on shore are scientificos. We will list those twelve as missing presumed dead until further notice. Details of our position, and list of missing personnel, will be enroute via normal data link in about five minutes."

The HQ Duty Officer came online: Sheldon repeated what he'd just told the HQ radioman to convey. Recognizing the Officer of the Day's voice, Sheldon then said "Hello Jim. Sheldon here". Then, grinning "Just wait until you see the maneuver we had to pull off to avoid being sunk by the ice! Half a dozen people got it on video. If it isn't on YouTube in the next hour, I'll eat my flat-hat." He glanced at Doc, who was the image of pure misery. "We had some amazing help from the civilians aboard – especially Doc – the whole hide-in-a-fjord idea was his. He's a damn good ocean-racing mentor, too. Wait a moment, somebody is waving for my attention. Break in transmission, Jayhawk back in a moment. Out."

SURVIVORS ASHORE?

"WHAT!?" asked the Captain, of the radioman who was, all of a sudden, frantically pointing at his own mouth and ear, obviously desperate for the Captain's attention.

"Sir – No trace of Teams A, B or C on either their normal voice-comm channel or the backup. But we checked the frequency for Team A's weather station transmitter. It's transmitting intermittently, switching on/off. Harry caught the pattern right away, he has a merit badge in Morse Code from the Boy Scouts thirty years ago. The weather giz is sending a perfect SOS, then the number four. Over and over again. Somebody's alive at that station, Sir! Team A has no gear with them that could automatically send something like that. Someone has to be doing it manually."

The Captain stared at him, then called Meadowlark. "Lark, this is the Captain. We have voice and data comm with CGHQ. Good job. We also have a repeating manually-keyed Morse signal from the Team A weather transmitter… reading S-O-S-4. We guess the "four" means all four members are okay. Hope we're right. You hightail it NOW over to Team A, and get them aboard. Abandon all of Team A's gear... but make sure they leave the weather station operating normally if possible. Might as well get something useful out of this exercise. But NO repeat NO delay to fix or adjust equipment. If the station isn't working when you arrive, forget it! Just bring the people home."

The Captain at once passed the word to both HQ and the entire ship – people who were just beginning to struggle out of emotional shock suddenly found themselves back-poundingly happy, the possibility of having four survivors being ever so much better than the near-certain presumption of zero. Even Doc's mood lightened perceptibly.

Meadowlark-2 cruised south towards Team A's site, flying close to the sheer outer faces of the coastal range. Curious, Marcia dropped down low, studied the amazing collection of floating broken ice. One did get the impression that the whole mass, horizon to the mountains, was moving, but much more sedately now, with none of its earlier fearsome churning. Piloting was straightforward, freeing a bit of her mind to wonder how deep the ocean-bottom was, along this side of Greenland – with so much ice floating about, there might be a problem clearing the stuff out the southern end of the current and into the open North Atlantic. What if the water were too shallow to pass the ice? What Jayhawk really did not need now was for Jayhawk's Fjord to get blocked forever by some goddamned chunk of ice that was too big to float or to be pushed from behind by the rest of the pack, and which chose to ground itself blocking the door. Because the current was actually carrying the ice southward, sooner or later the fjord's door would undoubtedly open again and Jayhawk could get out. How long it might be before that happened was clearly impossible to guess – but any data would help. Like, for instance, how much water this ice drew – how far down did it project – meaning, would it snag on the bottom anywhere soon?

Marcia flew Meadowlark-2 along the side of a large floe that had a nearly-flat vertical face, and told the crew "We need to know how deep the ice goes. Ice on the right – each of you give me an estimate of that vertical face's height. I know it's hard to do with nothing for scale reference, but try anyhow." She polled the crew: 140m, 200m, her own 225. Not bad agreement.

An electronic check on accuracy would be good: the smaller chunks of ice between the huge floes had to be floating with their upper surfaces pretty close to sea level, she thought: she flipped the height-finding radar on, brought Meadowlark even with the top of the face. Pinging off the chunky-stuff the radar read just under 200m above "ground".

"Good!" she thought. "Assume 225m of ice above water - that means about 8x225m submerged = 1800m to the bottom of the berg."

The ratio of ice draft to ocean depth would determine (a) if, and (b) most likely also when, the ice would let Jayhawk escape. The ocean in which Jayhawk floated was certainly not Marcia's professional field, but she was always curious about the geography of new duty-stations, and had studied the maritime charts of their work-area. She remembered the depth under this current to be about four or five thousand metres –so, this mess of ice right here WAS still afloat and going south, and presumably so was the remainder of the pack – sashaying slowly down from the north towards Jayhawk's Fjord. Hooray! The god-damned mess might just clear itself out, perhaps even fairly quickly… fingers crossed, knock on wood et al.

Approaching Team A's coordinates, Marcia took Meadowlark up to the needed altitude, called Jayhawk and explained her ice-observations and reasoning, got one question from the Captain: "See any open water between the big bits? You know, leads and open spaces large enough for Jayhawk? Preferably interconnected leads?"

The answer was "no", which of course disappointed everyone. But her analysis of the draft of the ice was met with an enthusiastic rush to find or download the best available bathymetric charts.

Marcia remembered quite clearly the shape of the mountaintop where she had set down Team A, and didn't need the GPS for the last few miles of approach. At two miles, her copilot peered through the big binoculars for a few seconds, then said, quietly, "Gottem!

Four dots, orange survival suits. They are going to be some happy campers!" Marcia relayed the information to Jayhawk, and through an open microphone on the bridge she caught just the first moments of cheering in the background. She grinned slightly to herself, then re-focused on the task at hand.

The mountaintop had changed shape: the nearly-level top had been about 20m x 40m, now reduced by half, but still usable. Rotor clearances were still good. One member of the Team used the proper hand-signals to help guide her down. Body language even at 20m separation and in survival work-suits was exquisitely clear. The leader, senior graduate student Charlie, trotted over, leaned in, shook James' hand –the only one he could reach- and motioned his team members to get aboard.

James handed the leader a helmet and headset, said "Glad to see you folks again. Query from the Captain – can you leave the weather station up and operating? If not, no biggie. We are to collect you folks, not salvage equipment. Doable in sixty seconds or less?"

Leader replied "Done already - as soon as we saw you. We figured that out ourselves. The system is properly installed and on the air right now, and should be good for two years. Despite us almost losing our whole damned helipad. Lost the radio in the ice collapse, that's why we had to send our SOS using the weather-station. We were a bit worried, you know – it seemed highly likely that the ice-cap's smashup might have swamped Jayhawk, and we had no receiver so we couldn't even ask. We got pretty damned lucky! Hell of a fine place, the A-Station, if your life goal is to starve or freeze to death."

Helmets all around, doors closed. Marcia lifted off, brought Meadowlark onto her return course, announced "Time to head for Jayhawk – she's trapped in her personal fjord for the moment, but undamaged… heck of a lot nicer place to vacation than your half helipad."

Charlie asked "Hooray for Jayhawk's being okay… but how the devil did you folks manage THAT?"

The story took only a minute. Then, with autopilot set, Marcia raised her faceplate, stared back over her shoulder briefly and said "OK, guys, anyone want to volunteer and tell us YOUR story?"

Charlie seemed to be the anointed reporter: "The collapse happened during sleeping-bag time, so we didn't see much. Big roar, huge sounds of something big and coming closer very fast, then the whole damn mountain shook for about five minutes: we had to stay on hands and knees, couldn't even stand up. When things calmed down, the radio and most of our other gear was gone. Lieutenant, you dropped us off, so you'll remember how the ice came right up to the edge of the flat top?"

Marcia nodded, said "Yep. That it did. But no more!"

"Yeah, no more, believe it!" said Charlie. "I'm the team geologist. Glaciers erode mountains by infiltrating water into crevices in the rocks, where it freezes. In freezing, the water expands and cracks the rock loose from the underlying material. The ice hangs onto the busted pieces. Then when the glacier flows downslope, that ice-grip on the underlying broken rock yanks the rock free and downslope it goes. What happened to our site was exactly that – when the ice left us, it took away half of our flat spot – as you noticed. Just plain dumb luck it didn't take every bit of it and us too."

Then, quietly but obviously for the entire A-Team, Charlie asked simply "B and C?"

There was nothing else to say, no other way to put it: Marcia shook her head and said "No word. All listed as missing, presumed dead. So were you four, up until about thirty minutes ago.'

AN AIRBORNE ANALYSIS

Then in answer to the query "What the hell happened? What do you know?" Marcia filled them in as best she could.

"At least this side of the Ice Cap seems to have slumped about a thousand vertical metres, cause unknown. But given my geological expertise, which is zero, I'll bet on structural problems stemming from warming. Foundation ice being melted, and so forth."

"For sure the slump goes as far north and south as we can see from a couple of miles altitude, and my guess is the slump goes the whole length of the island. All that ice has gone somewhere... maybe it stopped enroute to the west coast, but I wouldn't bet on it."

"Some fraction of the whole Cap has probably gone into the ocean – if it ALL goes, which obviously has not happened – at least, not yet, I've read that it would raise sea level by nearly eight metres worldwide – that's about twenty-six feet. Nothing quite like a couple of dozen feet of saltwater in every basement and subway station in Manhattan – with, just for argument's sake and based on what just happened, let's say one day's warning! At least for the moment it doesn't look THAT bad, but who can say... we probably need new satellite photos to be able to tell what all has happened. We've still got ice-cap-ice in view as far as we can see inland, so the very worst thing hasn't happened – at least not yet. Beyond visual range from ten thousand feet, who knows what's going on? At least, the remaining ice seems to be standing still for the moment, but again, who knows how long that'll last!"

"At any rate, there's no place for us humans in such a spectacle- we're way too soft and fragile. If the ice made it to the west coast, god knows how many would be killed."

ANALYSIS ABOARD JAYHAWK

Once Meadowlark's report was in hand, and Marcia had been debriefed by Rebecca and the Captain, Rebecca asked Captain Sheldon's okay to download up-to-date satellite views. Permission given - unlimited access to comm, phones, and to Jayhawk's not inconsiderable computing capability.

Rebecca's very first question was "How much of the Cap – what volume - has gone missing?" That was the critical core datum. It would be central to understanding and predicting the event's impact both on the world's ocean, and more directly on humans. She knew that even a quite-small percentage of the Cap's ice and water could easily have totally unpredictable –and dire- consequences.

Getting that number would involve some complex 3-dimensional analysis of sequential photos, using changes in contour lines to estimate the missing volume of ice. She drew up a very specific request for photos and their analysis, sent it off to the weather/climate data folks with a cover note expressing urgency.

Meanwhile, she and her scientific troops studied the latest non-stereo imagery. Given nothing more than Jayhawk's own experiences, some fraction of the Cap had definitely made it to the coast and into the ocean beyond. Satellite photos gave an overview – the entire green-belt along the western shore was covered deeply with shattered ice, stopped with its flow-patterns intact.

Everyone who looked at it understood – the entire human population of Greenland was gone.

The volumetric analysis done with stereo pairs of photographs produced the needed number – a mere two percent of the GIC's volume had left the island and landed in the ocean. Two percent seemed a quite small loss, but it was by no means a trivial amount of water... a 'mere' thirty thousand cubic kilometres... which was way, way lower than Rebecca had feared.

In fact, two percent of GIC's volume meant 2% of the rise calculated for total meltdown – that is, 2% of seven metres, for a rise in global sea-level of a 'mere' fourteen centimetres.

The photos also showed an overview of what had happened: The destruction of the Cap's supporting structure which began on the eastern side of Greenland had propagated like a shock-wave, all the way to the western edge of the Cap. The actual slumped ice from the east side hadn't traveled even halfway across the island: actual physical movement (meaning active flow) had been massive, but had been stopped about half-way across, presumably by some enormously strong bottom feature. The shock-wave of collapse had continued on its merry way, destroying mush of the west-side basement structure. It had crossed the remaining half of the island and reached the abrupt edge of the ice at the inland edge of the west-coast green-belt.

On the west coast, the effect was deadly: for the entire north-south edge of the Island, the westernmost several kilometres were snapped off and set free of the main Cap.

The released ice, a noodle-shaped strip hundreds of kilometres long, several kilometres wide and well over a kilometre thick, was mostly at an altitude of two or three kilometres. Gravity, with simple water as lubricant, swept the shattered two percent across the inhabitable zone and into the ocean... about fifteen times the volume of the Lake Missoula flood. And with similar, but much more severe, results.

At least she had the critical number for volume... now Rebecca could begin thinking about the oceanographic and social consequences of that number.

Section Two:
INDIA'S "MEAN-SEA-LEVEL" PEOPLE
a.k.a.
Just Fourteen Centimetres from Nuclear War

4

RAMDAAS INTRODUCED:
HIS EDUCATION

"MSL" internationally stands for "Mean Sea Level". "MSL People" live permanently at sea level plus 50cm to minus 10cm. Such peoples are legion around the world, living mostly on the deltas of large rivers. Their distribution and number have nothing whatever to do with national boundaries.

The so-called "Ganges River Delta" up in India's northeast corner, at the northern end of the Bay of Bengal, is created by the confluence of three major rivers – the Ganges (respectfully nicknamed "Gangaji"), the Brahmaputra and the Hugli. Politically, the 'Ganges Delta' is shared between India and Bangladesh. Living in/on the Delta are a great many MSL People: estimates range from 500k to over 2.5M. They assert no national identity. They subsist on fish and the vegetables grown in gardens atop the ever-shifting sandbanks and islands, most literally 'here today, gone tomorrow'. Their entire existence is driven by the rivers' water-level and its day-to-day changes.

Gangaji's "MSL territory" is only a fraction of the total Delta, but it is thickly packed with people, permanent life-long residents. Across that territory, major storms regularly bring three metres of water in from the sea as a storm surge, and the monsoons regularly provide a similar rise but from the opposite direction. All that extra water drains quickly, but often, meanwhile, there is, today, literally no place to stand, where yesterday MSL People were dense-packed. Given such spectacular flooding (often metres of rise in a period of hours, sometimes up to two or three or four times in a year), the MSL People have no choice but to move occasionally to higher ground, temporarily. Which leads to very sudden mass migrations of large numbers of quite desperate people who could not be clearer about their disdain for artificial borders. The lines which concern the MSL migrants are on topographic maps, not political. Importantly, the higher ground to which the migrants aspire is already densely populated as only India and perhaps China can do. "Upland" starts at about three metres above MSL – an altitude still subject to flooding but only during monstrous storms. Upland residents, along with their forebears, have owned and fully occupied every hectare of the Uplands for some thousands of years. For anything one can name, down even to mere standing room, Uplanders are forever in short supply. Uplanders have no interest whatever in sharing their own already seriously inadequate resources – not food, shelter, water or anything else.

{T_0 minus several years, continuing today}

Ramdaas, member in good standing of the Delta's MSL People, today aged 17, small-framed, boyishly good looking with a completely unruly mop of jet-black hair, brilliant white teeth that although perfect seem too large for his mouth when he smiles. Ridiculously skinny in that very special way peculiar to young Indian males of the lower castes. Which he is – he being a fisherman, he and his father and grandfathers for many misty, untold generations. That puts him in the Vaishya caste, third down from the top of the four major divisions of the Hindu caste system. His personal caste - as fisherman- is somewhere down close to the bottom of the intense pecking order within the Vaishyas.

MSL People are almost 100% illiterate – there is no hard-rock-based geology on the huge delta to physically support such permanent things as roads and schools. Most MSL individuals are worse off, educationally, than simply being illiterate: in point of fact, there is neither any cell-phone coverage, nor installed electricity. Batteries are both scarce and (relative to income) hugely expensive - as are today's small radios. Consequently even news –much less education- is in short and intermittent supply. Furthermore, it is clear that the local and national governments regard the MSL People as merely annoying, a nuisance but not a problem. To the fullest extent possible, the MSL People are simply ignored, and India can ignore people and problems with the very best! As can Bangladesh. The process of ignoring encompasses both voting, and the educating of children. Of which latter there are at least a plethora. No MSL person is known ever to have voted in an election. As with the national and regional governments, the rest of the world likewise careth not.

RAMDAAS'S EARLY EDUCATION

Ramdaas is brilliantly intelligent, and also almost charismatic – it has been obvious to his community for many years that there is something special about him. His history is complex (as, really, is the story of almost every person!). The first-born of three children, the only son. He began fishing with his father as soon as he could swim from boat to shore, say age four: by age six, although a small and rather un-muscled child, he was the fishing-strategist for Father – they were quite a successful team. The community knew of Ramdaas's uncanny abilities to understand and find fish, and were astounded.

Ramdaas's illiterate parents fully understood that education was the only ticket for escaping the family's poverty. On the high southern bank of the river –the India side- several difficult and unpredictable river-km away, there was a small public school, available not only free (for those who could get there on their own) but actually subsidized as to books and supplies, lunches and clothing. Politicians and officials cited the school's existence almost weekly to prove a lack of governmental neglect of regional social needs. His parents insisted that Ramdaas go to that school –beginning at age 6, the absolute minimum for entry. To make it possible, they refused to let him fish except on weekends, and managed to get for him a personal tiny sailing-skiff for the one-hour trip each way… as if to show off, he usually arrived home with a significant catch collected enroute.

Like his parents, Ramdaas's MSL People at large fully understood the value and usefulness of even minimal book education, but due to the intense poverty of the region, for a considerable radius he was the only youngster regularly attending school – rapidly if unintentionally becoming "THE" educated man for the whole area. He was quite extraordinarily good in mathematics and anything having to do with science, right from the start. Even before the end of first grade, before reaching age 7, and as soon as he could read at all, or add two-digit numbers, he became the unofficial record-keeper, resident calculator, scribe and reader for an ever-widening circle within his local MSL community. His willingness to help others was well known: often when he got home there would be a line of people waiting patiently. His mother had to be a sort of scheduling secretary. Ramdaas enjoyed the work - it exercised his brain mightily, the business of helping friends, neighbors, strangers to define problems and find ways through life's manifold difficulties, no two of which were ever identical. By age eight, he was often called upon to negotiate deals, and to adjudicate disagreements of all sorts. He had an eerie knack for seeing through the foliage to the actual crux of a situation. People did try, at first, to pay him for specific items of help, but he refused the offers: his parents agreed that it was good to use what the gods had given him for the betterment of their whole community – the family would themselves reap plenty of benefit without being unseemly about it.

His school was a frugal, waste-not institution, but sometimes he could find a cast-off newspaper, which he would bring home with him and read aloud to an admiring, attentive (and growing) group squatting at his feet – and then, almost always, he would find himself leading a discussion of some interesting or locally-important topic, being deferred to by almost everyone. He was innately both modest and honest - if he knew nothing helpful on a topic, he would say so, and promise to extract all the pertinent knowledge his teachers contained - "I just squeeze them like oranges, for their knowledge!" as he put it. He would always return next evening primed and loaded for an interesting discussion with an eager constituency. A truly voracious and omnivorous reader, his widening and deepening knowledge was an important ingredient in such events. As his abilities and experience increased, even in his earliest teens Ramdaas was becoming, unintentionally and steadily, a more and more important fixture in the local social and political landscape, a respected and trusted public figure. He never seemed to make a mis-step, even in the most complex of circumstances.

When his two sisters reached matriculation age, their parents insisted that women also must learn, at the very least, how to read and write and calculate – things the parents could not do for themselves and the lack of which persistently affected them. No way could all three sibs go off daily, for school, in Ramdaas's tiny boat. So the parents assigned the girls' education to Ramdaas. He being several years ahead of them, it was his duty to think back to his schooling at their ages, and teach THEM at the appropriate levels. He continually checked his approach by consulting with the lower-grade teachers. He quickly discovered that to firmly embed knowledge in your own brain, there is nothing like trying to teach another person something you think you clearly understand. Nor is there any better exercise in 'how to explain stuff to other people'. Effortlessly, he became both teacher and guru to the entire community.

Then, at age 14, one year before starting high-school, Ramdaas lost his mother to cholera, a not uncommon death given that the three delta-rivers are the ultimate sewage receptacle for tens if not hundreds of millions of humans and even larger numbers of livestock. Next year following, his father drowned in a sudden storm that killed more than 80 local boats and 140 men within a two-hour span. Most of the bodies, including Father, were never recovered.

With that savage storm, everything changed for the family. After the storm, Ramdaas had located his father's boat, repaired it, and set about being the breadwinner for himself and his sisters. That seemed to be the task which the gods had chosen for him, and he undertook it without rancor or upset. It meant, of course, quitting school, so as to fish full-time. That was too bad, indeed, but what could one do?

His needs, and his plan to drop out of school, were instantly known to his entire community. The decision to drop out caused an immense uproar, and triggered intense discussions across the community, which had not until then fully recognized either of two things – first, how much they had come to depend on his help in various ways, and second, how much of the community's self-respect (specifically, their between-groups bragging rights) was now wound up in their having Ramdaas as community figurehead - their very own unique reader-writer-teacher-mathematician-negotiator-arbiter.

Ramdaas had missed only two days of school before a delegation of MSL elders, half of them men, half women, approached him with a plan. He had done so much for others, it was repayment time. The community would draw from a pool of nearly 1000 local families to pay his living and school expenses to go back to school to enter and then finish high-school – the first ever locally! He must do so, for the honor of, and on behalf of, the MSL folk. And while he was in school, perhaps every weekend he could talk to a public group – to whoever might like to attend – explaining something from what he had learnt that week. After all, they had for years enjoyed having him explain things on a great many topics and scales, sometimes quite complex and mysterious things from his science courses, other times various current events. The whole village agreed that the tradition should not just continue but grow.

As this utterly unusual plan was being presented to Ramdaas, he was -slowly- realizing that he had inadvertently become the leader of something – not at all certain what it was, but he could sense the underlying latent power of all these MSL people thinking, discussing, working together to solve problems. And whatever this thing was, it was moving, growing, taking on tasks, seeking to accomplish things. It all sent shivers up his spine.

It was also during the presentation of this plan for his support that the delegation promoted him to the honorific title/position of Ramdaas-ji, shortened at once to Ramji. His protests, his claim not to in any way deserve the "title", availed him exactly nothing whatsoever. This, too, seemed to be something that the gods were doing to, or for, or with him: Ramji therefore accepted the accolade, and the overall plan.

{T_0 - 3.5 years}

RAMDAAS'S EDUCATION CONTINUES

Now, as he moved into high-school, Ramdaas encountered a first for him – a female teacher. Not just any teacher, but of math and science! And an extraordinarily good one – most teachers in semi-rural environments like this one had at best a bachelor's degree, and usually no training or experience in teaching. His teacher Dr Lakshmi actually held a doctorate in biological science, and an adjunct professorship at her husband's university nearby. In some areas she knew ever so much more about fishes and fisheries than did Ramdaas... but she was a very experienced, and genuinely GOOD, teacher. She would never ever show off her knowledge – rather, bit by careful bit, she would reveal it to her students so that they could make it their own and suddenly be her equal on that topic.

He was envious of her education, vowed never ever to stop learning. He devoured anything about math or science: the concept of 'ecosystem' especially entranced him – studying all the bits of the universe that had to work together, be they physical or biological or chemical, for any living system to survive. By the end of year one of high-school, he was completely hooked – the world was so complicated, so difficult to understand, and so beautiful! In turn, Lakshmi was fascinated by his intellectual voraciousness, his inherent ability to integrate disparate bits into a coherent whole. Ramji read science as if starved – but at best the available information in the school's meager library was some years old and hardly very technical. The school's materials were augmented as much as possible by journals, books, articles, even government pamphlets and planning documents, all funneled to him by Lakshmi and her oceanographer-husband-professor, Dr Hamid, the materials coming mostly from the university library. Largely on Ramdaas's behalf she visited the university library almost weekly, a fifty kilometres roundtrip. The thick, well-illustrated books which she and Hamid provided, and which Ramdaas regularly brought home to study, were to all MSL People who saw them objects of fascination– almost, in fact, of veneration. He liked showing the materials around, and consequently spent a great deal of time explaining graphs and pictures.

{T_0 - 2.5 years}

RAMDAAS'S EDUCATION BROADENS

It was in his junior year that he discovered physical oceanography – Lakshmi's husband came to the school once or twice per year (at his wife's invitation) to talk to the advanced students in the hard sciences. Although formally a physical oceanographer, Hamid's main interest, as well as his career, was in global warming. In their discussions, Ramji was stunned at the idea that mere humans, however numerous and whatever they might be doing, could actually physically change the planet and how life on it behaved. And almost certainly for the worse! Many people, he knew, chose to deny or ignore the process. Yet the evidence shown by and discussed with Dr Hamid was so CLEAR! Worrisomely so.

Ramji's math helped with the physical oceanography and planetary atmospherics that Dr Lakshmi and Dr Hamid threw at the class. He understood tides almost instinctively... the nearly-perfect predictability of the solar and lunar tides, the random huge chaotic fluctuations due to storms, which were impressed upon the underlying regularity. All this made perfect sense to him, a deeply personal form of knowledge, given how his entire community interacted with the intertidal environment. And the connection between intertidal zones and the results of planetary warming were impossible to miss, a topic of intense discussion with Dr Hamid whenever possible.

It was clear as a bell to Ramji that global warming existed, that it was going to continue, that the warming would inevitably melt some or all of Earth's ice, and that the meltwater would raise sea level quite significantly. Which would certainly affect the intertidal zones near MSL, and would equally surely affect how the MSL People interact with tides and storm surges. And one could predict with perfect certainty the inevitable, catastrophic effects of rising sea-level on his own MSL People.

It seemed obvious that only planning and resources on the national-government level and above could address such a problem. Therefore he worried his way through the government materials – found himself continually dismayed - such a high percentage of garbage in there, passing for either science or simply information! Why could he, a low-caste orphaned 16 year old still in basic school, see things so clearly, see the garbage for what it was, whilst these highly paid highly educated government folks could NOT? India's government couldn't seem to talk about the problem in a time frame of greater than perhaps ten years: Ramji's basic "ECOLOGY-101" level of ecological training enabled him – no, better to say it demanded of him - to think in terms of centuries and (preferably) millennia. After all, this was INDIA for the Gods' sake, with its 5000 years of written history! Who else on the planet should be better qualified to think in the long-term?

Ramji was in turmoil trying to integrate everything he was learning with how the MSL People had to live – and in particular he worried about the ramifications of warming, and rising sea-level. He didn't know it but he was innately politically astute – he understood from personal experience in the MSL flood-avoidance migrations that the Uplanders possessed and would kill to retain the only escape grounds for MSL People beset by storm or tide – and he understood quite clearly the implications of changes in sea level for the interactions between the two groups. Without some sort of serious, careful planning and forethought, the MSL migrations could only become more intense and more prolonged and more frequent, more dangerous and disruptive.

If MSL migrations were inevitable (and they were vitally necessary, simply because no human can stand in place in water three metres deep and moving at several kmph) then so was conflict, better called what it would be, namely war. Absent major changes, there would eventually -inevitably- be war between MSL People and Uplanders – perhaps at the technological level of sharpened bamboo poles if the two nations continued to ignore the situation, perhaps at god only knew what escalated level if the countries actually chose to pay attention. But intelligent well-meaning people should be able to see problems coming and prevent them. Only the Gods' own behavior was beyond human control, and people fighting people was human behavior.

It wasn't clear to Ramji which alternative might be preferable, ongoing neglect or increased attention. The BEST future would of course involve some sort of coherent sensible two-nation coordinated plan between India and Bangladesh. But given that the national boundaries involved were some of the most highly artificial and genuinely idiotic ever drawn upon parchment by human hands, the chances of such an international, intergovernmental plan, he understood, were those of a snowball in hell.

Fortunately, any sea-level changes, any significant variation in the tides, seemed to be some years off – in eco-time those would be serious problems indeed, but in local-political time – with its five-year horizon - nothing much seemed to be happening. After all, a few mm or cm of rise in MSL was of immediate importance only on some mid-Pacific atolls that should never have been inhabited anyhow, and simply did not grab the public's, or its government's, interest.

{T_0 - one year}

Meanwhile, Ramji had tasks to perform – his weekly almost-a-seminar to the village, covering items from what he'd learned that week. Getting on through school. Teaching his siblings. There was in his expanding sphere of knowledge a slowly-growing worry about the things he was teaching, the warming and the sea-level rising and all that – no good whatever could come of such "humans-playing-god" games. But nobody had yet seen anything overt that was attributable to those mysterious happenings: all negative effects appeared to be so far off in both space and time that ignoring the problems seemed entirely rational. There was just a vague but persistent feeling that storms were getting stronger and more frequent.

Ramji's weekly discussions of what he was learning continued to be evening entertainment for his and several other villages. Attendance grew steadily: topics ranged widely, but a frequent theme for discussion was global warming and especially how much and how fast the water level in the Delta might change. The entire concept made most people extremely anxious... having to really worry about the unknown, or the 'known-but-not-well-understood', was no fun at all.

{T_0 - several months}
RAMDAAS'S PROFESSORS VISIT HIS "CLASS"

With considerable nervousness, and only after a great deal of screwing-up of his courage, Ramji invited Lakshmi and Hamid to come visit, to attend and participate in one of his "town circle discussions". It would mean a two-way voyage of several miles in an open fishing-boat, hardly the cleanest or most comfortable of conveyances: it was only then that Ramji discovered that neither of his teachers had ever been out in a boat on the river, although both – thank the gods!- claimed to be able to swim reasonably well.

The single thing most worrisome to Ramji was not introducing them to his People, or the boat trip – no, it was the prospect of inviting his guests for dinner at his own home, something that simply had to happen given the timing of the visit and class, but it would be a cross-caste invitation that in the days before the official end of castes could not even have been contemplated - after all, his teachers were both Brahmins, stratospherically higher than Ramji's station. He explained his nervousness to them, ended by formally extending the invitation. To his relief, both teachers pooh-poohed the caste "problem". To his delight, the pair instantly agreed to the whole package with unfeigned enthusiasm, especially when they learned the meal would be prepared by Ramji's two sisters, whom for several years now they had been helping Ramji to educate, and whom neither teacher had ever met. After dinner, the guests would be returned to the schoolhouse well before midnight, via boat. If they could arrive that day at school in their car, it would be a wonderful help because the couple could then get themselves home safely from the school, despite the inconvenient hour – a time of day with a guaranteed dearth of available taxis.

Picking a date was easy - Wednesday would be best, because on every Wednesday classes ended two hours early. Perhaps Wednesday two weeks hence? All agreed.

Then Hamid asked, "Seriously now, Ramji, what is the attendance likely to be? It does help a teacher to know such things!" Ramji grinned, embarrassed - "My sisters counted the audience two weeks ago. As you taught me, I had them count independently, they of course got slightly different numbers but the mean was 956, and the girls agree that the crowd seemed of ordinary size." Then after a pause to study the teachers' astounded faces (from Ramji's modest accounts, they had had no idea how big Ramji's "classes" were, would have guessed a couple dozen attendees at most) he said "I will announce your acceptance of our invitation – I say "our invitation" because, believe me, the invitation comes not just from me personally but from my entire People. I would expect that participation to double. Perhaps more. After all, my People have heard a great deal about you two!"

That evening, the announcement of the teachers' impending visit created an enormous stir: nothing like this had ever happened before, the village was instantly thrown into a tizzy of anticipation and pride. Ramji's local fame and credibility, already high, soared. The largest fishing boat in the village belonged to Three Brothers: they immediately volunteered the craft and their services to transport the visitors. When Ramji accepted the offer, all of the village's fishermen turned out to scrub and repaint the boat. After which it was doused in rosewater and repeatedly fumigated with incense-smoke.

The week before the visit, Lakshmi took Ramji aside and said "If you actually have an audience of two thousand, we cannot possibly reach them all with just our natural voices. I will bring the school's battery-powered portable public address system. Saves the vocal cords, makes the listener's job easier." The three then planned the topic - a review of the global warming problems that –as a result of Ramji's teachings- so concerned and deeply frightened all MSL People. The three teachers had no idea how far off that particular track their meeting was going to stray.

5

"MIGRATION PROGRAM" CONCEIVED

Came the chosen Wednesday. One Brother steered, two rowed, the guests sat in the stern facing both the oarsmen and the destination. To their great embarrassment and utter delight, the Brothers had been individually introduced to Lakshmi and Hamid, as they boarded. On the trip, Ramji played tour guide, discussing the river, its behavior, fishing and fisheries, and especially the need to make occasional whole-population migrations to avoid floods. The Brothers listened like bats - before the one-hour trip was fifteen minutes old, they had added comments to things Ramji said, and suddenly they found themselves in actual conversation with the two exalted personages - and as equals! Unbelievable! Ramji was diplomatic in the extreme, handing to the Brothers many of the questions posed to him by the guests. Once over their initial brief reticence and self-deprecation, and as soon as it was clear that the guests respected their knowledge, the Brothers were astounded to find that THEY could answer questions posed directly to them by the guests - they themselves, mere fishermen, knew a great deal that these teachers did not - and the teachers seemed eager to learn from them. A most amazing and perplexing, but pleasant, state of affairs!

At about one kilometres offshore from Ramji's village, Hamid stopped talking, peered intensely forward, then pointed and asked Ramji "There is a tall pole, maybe five or six metres and painted white, standing on the bank – is that our destination?"

Ramji: "Yes, it is – we put lights on it so that our fishermen can find their way home at night."

Hamid stared again at the approaching bank, then asked "Ramji, the top of the sandbank is WHITE for about 100 metres to either side of the landing pole. Certainly it is not snow: what is it?"

Ramji looked briefly over his shoulder, turned back to Hamid and Lakshmi, smiled with delight, said "I warned you that your visit would be popular. That white atop the sandbank is just part of your class for this evening – we decided to have everyone wear white because it is the only color that everyone in the village is certain to possess. It is also intended as a mark of respect. Although my People do not yet know either of you personally, they feel that both of you are already a part of our local society. They are very proud of having a connection to you two!"

Then, with another broad grin, Ramji reached into the small backpack he carried everywhere, pulled out something neither guest had ever seen – a classic Gandhi/Nehru cap, but in flaming scarlet rather than traditional white or khaki. "A small group of my regular students made this and gave it to me – they say they want everyone to be able to tell from any distance just who the teacher is, if perhaps the crowd might someday be so large. The scarlet is supposed to represent my intensity when I speak to the meetings. Or so they say!" He turned to face the landing-site, waved, got an eerily-silent 2000-arm wave in return. Lakshmi cocked her head quizzically towards Ramji, who explained - "The peoples' silence is a requirement for such a large group – everyone here understands that. It is also another mark of their respect – they would not interrupt us, and they all, every last one, would hear every syllable of what we say. Only audience silence can ensure those things. But they do not yet know we have the PA: a nice surprise for them. So much less strain on both vocal cords and ears!" He paused: "We will provide lots of time for questions, too. After all, much of this audience has had years of weekly practice for just such an event as this. There are many, many intelligent people, potential good students, in this and every other village, if only we had a school! Such a waste this way!"

The boat slowed for the final fifty metres, headed straight for the pole, where a notch had been cut in the very low bank to create a good docking space. Lakshmi and Hamid simply stared – for at least a 50 metre radius, the ground was quite literally covered with squatting people... and there were many hundreds more standing on the outskirts. The boat grounded: in two seconds a gangplank was dropped into place, but there was no open path up the bank through the dense-packed crowd. Ramji pointed to where it was needed, waved his hand. Like magic, that part of the crowd stood as one person, and simply parted to clear their way. As Lakshmi and Hamid touched sand, they formally greeted the massed villagers with the classic 'Namaste' gesture, the anjali mudra, made to each cardinal point so as to include everyone. It was returned enthusiastically but silently by the entire crowd.

In the open central clearing the village had built a raised bamboo stage, about a metre off the ground, with plenty of room for the three teachers. The crowd was clearly bigger than even Ramji had expected. One of the Brothers brought the little PA system, and Hamid ran through the twenty-second setup, tested the system by saying "Hello" to the crowd, who answered almost silently again, by Namaste.

The lecture went extremely well: because there were many brand-new 'students' in the audience, the teachers undertook a quick foundation-up explanation of warming and its likely ramifications locally – those in the audience nodding rapidly were obviously regular attendees at Ramji's seminars... for them, this was mostly a review. The three teachers read their audience confidently, handled all the sticky points with clarity and humor, skillfully handing topics back and forth amongst themselves, changing the lead frequently. The audience was initially reserved and quiet – quite different from the norm for Ramji's solo classes: after some months or years of attendance, all of Ramji's students knew they could ask any question they wished in class and be taken quite seriously regardless of their personal background or knowledge... here, the newbies were still figuring that out.

The initial reticence of this huge audience was expectable, given the social rank of the guest teachers... not to even mention that one of them was FEMALE. The crowd had known in advance about Lakshmi's gender, of course, but were a bit bedazzled by confronting -and talking directly with- the reality. After all, the MSL People had been almost totally isolated from the last fifty years of rapid social change, and any active leadership roles for women were at the very least a topic for considerable discussion, if not frank disbelief. Lakshmi's participation had brought out probably ten times the otherwise-likely number of women.

By perhaps forty minutes or so, several questions had been asked by long-term students, and that had broken the ice for the entire audience. Ramji had been working with many of the more intelligent, aggressively-interested villagers for a long time, in fact for some years. Now, a pair of those long-term students undertook what they later admitted –freely- was a plan to change the direction of the conversation and lecture. Well into the general discussion, one of them, named Pranav -a frequent and serious attendee at Ramji's class- raised his hand, stood when recognized, asked "Ramji, Honored Doctors, my question has many sides to it, but I must begin simply as Ramji has taught us. First part of the question: is it guaranteed by the unchangeable nature of things that we MSL People will forever continue to have to migrate to higher ground almost every year, to avoid the floods?"

Hamid answered; "In the old way of speaking I would say 'It is so written by the gods and nature'. The answer is YES. The river will continue to flood from both directions as it has for thousands of years. There is no reason whatever to expect any change in that. But at least we understand it quite well today."

Pranav nodded, said in a strong voice, clearly addressed to the crowd as well as the stage - "Cannot such repeated disasters be stopped by building a dam or changing the course of a river? We know such things have certainly been done elsewhere - we have talked about them with Ramji."

Ramji spoke: "Yes, we have talked about such things. Doctor Hamid is quite correct, the river will continue its behavior, but in theory, if we humans wanted to do so we could build things to slow or stop or control those floods – they would have to be huge and expensive engineering works, but such projects are definitely possible. Even for an area so large as Gangaji's delta they are possible. But very unlikely indeed. After all, we have neither money nor influence!"

Student Pranav shrugged, then said with a tinge of bitterness, "Truly unlikely! Such things do get built elsewhere, but not HERE: unfortunately we MSL People count for nothing, so far as I can see." He paused, then said "Perhaps, honored teachers, it is up to us, ourselves, the migrators, to do something on our own? We cannot build a dam or move a river, nor can we control rain and storms. But because we know for certain that floods will come, and will require us to escape briefly to Upland areas, perhaps we could plan HOW to do such moves better. We sometimes have to migrate more than once in a year, and only seldom is there no flooding in an entire year. The need to migrate due to floods seems inevitable. Would it not be intelligent for us to plan for moving, to make a plan by working, all of us, together? To plan in some way so that fewer people on both sides, Uplander and ourselves, get hurt or angry or have their property damaged?"

The three teachers looked at one another, quite surprised at this new direction: Ramji recognized the structure of what Pranav was saying – it echoed some intense discussions in class a few weeks back. Before the teachers could formulate a response, a second man in the audience stood, raised his arm for attention. Ramji pointed to him, said "You, Amit! I have seen you many times before, in our weekly class. What would you say? It is your turn now."

Amit nodded to the teachers, said "That is a fine idea from my friend Sri Pranav, but unfortunately also this particular idea is not so intelligent as his normally are. All we have is PEOPLE. We have no other resources, no money and certainly no influence with our government. We as an entire people have no cash income to speak of so of course we do not pay taxes. Hence we do not exist! Nobody for time immemorial has cared to help us in any way other than to say "Good Riddance" when a flood carries off ten thousand of us who cannot get to higher ground for a couple of days."

Lakshmi thought to herself that this new tangent might be quite interesting, and determined instantly to pursue it. She raised a hand to speak – the crowd went entirely silent – few indeed had ever heard a woman speak out in public, as a leader! "I understand the comment about having nothing, having no resources… but that is not correct. If we are to talk about MSL People having or not having resources, we must think very widely, not narrowly. In fact MSL People have a great deal of the most valuable and important and strongest resource – you as a People have half a million human brains. I believe that such a collection of human brains working well together could solve any problem, of any sort, anywhere in the universe. You also have other resources – for instance, you have a great deal of time in which to think – time in which to make intelligent plans. You also know when and how the river will behave – not precisely, but with great certainty. Such knowledge is itself a resource. Knowing what one must plan for makes the task much more do-able."

She waved at Pranav, the first speaker: "So, Sri Pranav, your idea is not in any way a bad one… quite the contrary! It simply needs a great deal of discussion and thought before it can be intelligently considered… only after such a discussion might one think about a plan of action. But understand me clearly everyone --- I mean a REAL plan, with carefully specified goals so that one knows at every step where one is trying to go, and what one is trying to do. An example - 'Save myself from drowning during the next flood' is a noble idea or goal, but too broad. 'Save myself from drowning by having a boat when the next flood comes, a boat which I will build myself before I need it' – That last is a plan, the first is merely a dream." She shrugged: "The reality, unfortunately, is that dreamers drown where planners survive and even prosper."

After a ten-second private confab with his teaching partners, Hamid took up the microphone: "If it is acceptable to the audience, we would like to pursue this idea – we think it is much more important than continuing our original discussion of warming." He looked at Ramji and Lakshmi, got nods from both, and a smattering of affirmative noises from the crowd. A different tack indeed, and a most interesting one.

Hamid then spoke to Pranav: "Perhaps you have been thinking about this for some time?" The man looked embarrassed but nodded yes. "Then," said Hamid, "…have you thought about making a clear statement of the problem? Often a clear statement of one's problem opens unseen pathways. Any of the three of us up here on the platform can help you if you wish." Here he waved his hand to indicate the entire assemblage. "…we three can help all of you think through to the core of the problem. This is just a more complicated problem than say fixing a leak in a boat, but the way to go about it is largely the same. Break it down into tiny steps, taken in the proper sequence."

Hamid paused to re-gather everyone's attention: "And who knows, if you could come up with a plan to handle a carefully-described problem, perhaps the three of us might take a document containing such matter to the correct authorities. Perhaps your voice has never been made loud enough, or heard in the right places? Perhaps the three of us speaking together on your behalf might just have an effect… but ONLY IF there is a good plan, well presented. That means a plan you people came up with by yourselves, not a plan invented by outside amateurs like me and my wife!"

He paused, spoke privately to Lakshmi: "Tell me clearly – am I getting both of us in too deeply in this business? SOMEONE from the outside world must help these people!" Lakshmi smiled, patted his arm, said "Not at all. I think this is something we are both called to do. As is Ramji. Please continue!"

A few moments of silence from the crowd, broken by various expressions of amazement, amusement, bewilderment. Then Hamid said "You, Pranav, were the first to speak – what do you think the problem is? You know my teaching style - make it as simple as possible without error! Go slowly, think hard before you say anything!"

Pranav was silent for a long moment, thinking. The crowd waited, curious. Finally he said "The problem is that whenever the river floods, we must all of us migrate temporarily to high ground to survive. We cannot change that need, but even though the problem keeps coming back, there is no plan for having so many people all move at once. Every time, there is great confusion and damage and suffering. It apparently has been so forever."

Hamid nodded: Lakshmi asked "Sri Pranav - most likely you have been thinking about the whole problem, all aspects at once. But surely it must be possible to think of smaller problems each of which is part of the big problem? I suspect everyone here has made such a migration and probably much more than once, true?"

A scattering of agreements, some quite emphatic. Pranav nodded and said "Myself, many times."

Lakshmi asked him "Tell us all, from your own experience, what are some of the smaller problems that reside within the larger?"

He shrugged: "Doctor Lakshmi, simply the number of people is itself a problem, so many, a huge number, more than the stars but fewer than the sand-grains. So are food, water, moving the old and the sick and injured. Almost too many problems to count."

When he paused, a voice from the crowd added "Roads – or at least paths marked so we know where to go. And we always need shelter for a few days."

"Freedom to travel by way of specific pathways, known in advance" someone called out, immediately answered by an old man: "MONEY speaks loudly! If we could negotiate for the right to use particular pathways for some trivial payment, say one pice per migrator, that would give the land-owners a bonus which even WE could afford! The Uplanders would all be eager to have us instead of being upset."

Yet another voice: "Fuel to cook, blankets to be warm. Places of sanitation."

A fourth: "Just finding enough empty space for us to stay out of the water for a few days. But getting that much space almost always causes fights between us and the Uplanders – they don't even want us standing on their land, much less staying long enough to be safe from a flood. But even a tiny bit of money would surely change that!"

There was suddenly quite a strong and growing murmuring from the crowd, as neighbors tried out on one another some idea or reaction.

Now Ramji took off his red cap, waved it overhead at the crowd for silence. "Very good! A big problem made into smaller pieces might become something one can solve. For instance, think of sanitation. We all know what pit-toilets are, we use them everywhere. They are easy to dig, easy to cover and leave no traces. A small group who knew where people would be walking, and more or less how many people, could go ahead of the crowd and dig them in advance. That's a much less complicated thing than, say, building a good boat!" He paused, clapped his cap back on his head. "Here in this group are lakhs, even crores of years' experience solving problems of all kinds. That is a huge amount of knowledge. Think now – everyone! Consider another smaller "problem-inside-the-problem". What might we do about space to sit for a few days, and about shelter? Does anyone have thoughts about those smaller problems? Each is big in itself, but each is still only part of the overall problem."

An older woman in the far edge of the crowd stared hard at Lakshmi for several seconds, pumped up her courage and stood, to the utter amazement of most of the attendees. She raised her arm, was so obvious that Lakshmi immediately called on her: "Shrimati! Your name, please?"

The woman was clearly startled, unused to being addressed as shrimati, the feminine equivalent of the basic "sri" used in addressing men. After a moment to get going, out came her barely audible reply: "Madhu, Shrimati Lakshmi. My name is Madhu."

Lakshmi smiled, urged the woman onwards – "So, Shrimati Madhu, what would you say to us all about this big problem, or about some smaller part of the bigger one? Speak freely, speak directly to me, please. Just ignore the crowd."

The woman took several tries at getting started, finally locked her gaze on Lakshmi and blurted out in a rush "My thoughts, Teacher Lakshmi, are about one bit of the big problem. Back when I was just a girl, a long long time ago, I lived about ten kilometers farther upland. There are some very strange things up there. On the big flatlands, there is a tall metal fence that goes for many kilometres. Inside the fence is an old airfield, with many kilometres of wide, straight concrete roads for the airplanes, but there are no airplanes at all these days. None at all that I know of for the past many years, not a one. Yet there are many many huge old metal buildings, the houses for airplanes I believe. So many of them, the buildings! Lines of them, each bigger than any ship or boat anyone in this group has ever seen. And nothing inside – you can sneak up to the doors and look in through the cracks. All empty, all the time. Well made and well preserved, the buildings are - doors open easily, the windows still have their glass. I think the land and buildings all belong to the government, to the air force probably, but I have never seen any sign of them using the space or the buildings. They just have a few people, soldiers, who drive around the fence-line every day making sure nobody gets in, I guess. Guarding something nobody uses."

"If they would open the fence-gates and let us inside when the floods come, there is plenty of room for us all – just on the plain dirt and unused concrete. At least, I think so. We would not be bothering the Uplanders that way, especially if we marked clear paths for our people to use. And if whoever owns the airfield would just open the doors to the big airplane houses, I think there might be enough room for everyone to get out of the rain safely… not warm, but at least not wet. There are SO MANY empty airplane houses! And so much empty unused land! Such a waste."

Then, finding herself suddenly intensely shy after her long soliloquy, Madhu stopped, grinned in embarrassment, and ended with "Doctor Lakshmi, Doctor Hamid, Ramji – solving this big problem sounds very like cooking a dinner, lots of particular things to do in the right order, each small thing being important itself, but always with an overall plan. Every woman here knows how to do THAT.

And the men, most of them, know how to build or repair a boat. We have lots of practice making a bunch of small problems out of one bigger problem… it's just that before now, we haven't thought that way about **this** problem." Then with almost a giggle, "And as for another of the small-problems, well, an airbase must have been built to have lots of people living there, so there must be a water supply. Maybe with some good luck we wouldn't have to worry about THAT?" She paused again, took a deep breath, and finished with "Carrying water for several lakhs of people by hand, from river to the airfield, is not a task to please anyone assigned to it! I could go check on that little problem myself, to begin with."

She sat down.

Silence.

Hamid took the microphone, said quietly "Shrimati Madhu, you have a remarkable brain between your ears. Very quick, very sharp. I am impressed – no, we are all three of us impressed. Thank you for your ideas - they are wonderful! Exactly the sort of thinking we need."

Madhu, widowed housewife and illiterate mother of six (two surviving), had never been publicly praised before, for anything save running speed back at age ten. She was embarrassed into total fluster, but received friendly, supportive, admiring pats and words from her female neighbors on the sand.

Lakshmi spoke into the moments immediately following that praise: "I did not really understand the migration problem before we arrived here for this occasion." She took a deep breath, continued: "I believe that in such a group as this one there must be a huge amount of talent. Shall we begin here and now to figure out a solution to the BIG problem by breaking it into smaller pieces, then take the assembled overall solution to where the power and resources reside? You already have a leader…"

Here she reached out, took Ramji's hand, held it up in the air as if he were a victorious boxer. He stared at her, then went beet-root under his Endo-Dravidian base coloring as a huge wave of applause rose from the audience. "We will help, Ramji can lead. We may have enough people here right now, but perhaps we should call for a much larger assembly, since the plan should include as many of the MSL People as possible. What do you all think about that? The bigger our numbers, the better for making arguments and having discussions with authorities!"

As she spoke, Ramji's brain was whirling, he was coming up to speed remarkably well – and one tiny corner of his mind realized that right NOW was the pivot-point of his life. He would think about that – sometime, preferably somewhat later! Meanwhile – quickly into realities. An even bigger meeting? Not a good idea until the problem had been carefully discussed SOMEWHERE, because a really big meeting without a controlled agenda would be instant chaos, counterproductive in the extreme. Within the present group would be a fine place to start. Most of this assemblage knew one another, or at least were neighbors. They could take the problem-bits already identified (his mind bifurcated for a moment, produced a list - numbers of people, sanitation, food, shelter, roads, law-and-order, space, routes, fuel, water, food), find the right people to work each one, massage the result into an agenda that would support a much larger meeting. That is, IF they still felt such a meeting was needed. Then massage it again to suit political and administrative and financial needs of all concerned – a very complex undertaking. For instance, it would undoubtedly become a rather sensitive task just figuring out who (whether bureaucratic agency or individual human) would end up in the "all concerned" list.

Ramji took the microphone, called for quiet. "Let us begin right here, with our own villages – then bigger meetings later if they are needed. We all know there will almost certainly be a need for a migration in, let us say, the next year. If we start IMMEDIATELY, we might be able to have at least a good preliminary plan for some experiments during the very next flood. The right experiments would teach us all a great deal, and that would let us better handle the next migration beyond that one. We will learn a lot from thinking about the problems and also from trying to actually do something. Then our NEXT time should be much better. Sort of like learning to swim, I think! I suggest we do this – take all the topics we discussed, the smaller problems, and begin to think right now about the talents and abilities of your friends and relatives and neighbors – then tell us, tell the whole group, who from among you should be in charge of looking into each smaller bit – someone who can get things done, who is smart and can think clearly and is not afraid of new ideas. Shall we begin here and now?"

There was a mild roar of approval, and out in the middle of it, standing side by side, were Pranav and Amit, grinning at one another – things had gone so much better and so much farther than their most optimistic imaginings, so fast that they could hardly believe it. Each had come prepared, hoping: now, each whipped out his own scarlet Nehru cap and donned it. There was a sudden silence as a growing circle of people around the two caps fell silent, wondering.

The three on the podium saw the action: Hamid looked at Lakshmi, whispered "I have such a feeling as never before in this life! What we do here and now will go down as a truly historic moment – and we get to participate." She replied "Husband, we may well have here another Gandhi, our very own. You and I seem to have been chosen by the gods for a very special role in this young man's life. We must rise to the call!"

Over the PA, Ramji called once more for quiet. His brain was racing, but it felt as if he'd rehearsed all this forever. "Each small problem needs its own leader. We will shortly decide who those leaders are. But I would choose the first of them immediately. We have already seen the problems of 'space' and 'shelter' addressed very clearly. Shrimati Madhu, please come to the podium with us."

Madhu tried to protest, to refuse: the women nearest her wouldn't hear of it, almost bodily propelled her forward until suddenly she was on the podium with the three teachers. Ramji took her gently by the elbow (he was small, she much smaller yet), turned her to face the main audience and said into the microphone "This is a new age and this is a huge and difficult problem. We will NONE OF US care at all in what bodies the necessary minds reside. We will be concerned ONLY with talent, with getting the task accomplished. No other behavior is acceptable, because only with the very best thinking can we hope to succeed."

He understood intuitively the politics of things, symbology, how best to behave. He stepped back a half pace, reached up, took his cap off, and placed it neatly on Madhu's head: she gawped, surprised into silence and immobility – women did not wear such a thing at all, not ever! Before she could do or say anything, Ramji went on - "Such a red cap will be worn by each problem-leader. Here is the very first – Madhu, your task right now is to find people to help you – we need to know WHAT is needed and HOW MUCH of each thing is needed, and also how much might be available if we can shake the government's mango-tree. Will you do that? Choose only a few, the very best people, do not be in a hurry, do not let friendship or relationships get in the way. Best talent only. Because you and your group are our very first, you must become our example."

Ramji's little speech had given her time to recover. She reached up and patted the cap, proudly settling it more tightly, grinned shyly at him, then at Lakshmi –who was beaming- and Hamid.

"I will do it."

Ramji looked out into the audience at Pranav and Amit, called them out by name: "I believe you two had something to do with this --- so you can each lead a little-bit problem … but you must also find about ten more of these red caps! Pranav, you are famous for your love of arithmetic, so your first task can be to get a count of how many people will be moving – it must be a reasonably accurate number, because it will become the basis for all other planning. Don't try to be perfect, that is impossible… but do try to be thorough and always do the counting the same way so that counts taken at different times can be compared to one another. Find helpers, bring us a plan for getting that number quickly!"

Hamid stopped a grin, bent down and said softly to Lakshmi "Just LISTEN to him! He's using things we helped him learn… and explaining them well, too. And to this audience! Amazing!" Lakshmi just nodded happily.

The class had been planned to run for two hours only – it ended after nearly four, the last two being a very large and free-form "committee of the whole". From the apparent chaos and turmoil emerged two lists agreed upon by all: the first identified important sub-problems; the second, individuals and pairs somehow suited to initial consideration of a particular "little bit" sub-problem.

They adjourned until the time for Ramji's class next week: that time being already well established as Ramji's own, it would become the regular business meeting. The actual class would also continue, but be moved to another day.

TEST ASSEMBLAGE

It turned out that to properly advertise and then organize the Migration Project (quickly and permanently nicknamed "MP"), several very large general-public meetings were necessary. Such a concentration of humans on a small area of the open Delta could go unseen by the authorities once or twice, but not always - at least one such huge early gathering was caught on aerial-drone film. That resulted in a return photo mission two days later – by which time the entire assemblage had dispersed, gone home. The implied level of organization drove the intelligence folks crazy – but not nearly so crazy as did their speculation as to WHY the crowd had assembled. Insurrection, revolution, uprising… they were certain it must be some such thing driving the meetings! The intelligence people were most disturbed because they could not get any good, hard information: the MSL People form a closed and remote society, nearly impossible to infiltrate. The intelligence folks' antennae went up, picked up rumors about Ramdaas and the MP – all of which was thoroughly misinterpreted by being run through various paranoia-adders. Total misunderstanding of the phenomenon resulted in lateral transfer of information at the lowest and most paranoid levels of the three governments – but not the expected vertical transmission, up the proper channels.

The search for 'RedCaps", leaders to head groups working on sub-problems, produced an amazing array of qualified and highly capable people, all interested in this process, a project that would be orders of magnitude larger than anything they had ever even thought about, much less tried to do. At once, Ramdaas added a second series of weekly meetings, the "RedCap meetings" for MP group leaders. Each group prepared a plan of action for working on their particular problem, figured out how to define that problem's needs, where to look to fill those needs, likely stumbling-blocks, necessary coordination with other groups.

The participants were all pragmatists, all knew one another at some level, were actually quite used to communal problem-solving. Their illiteracy didn't get in the way of their ability to think, and to observe – when Madhu's idea of using the airfield as a refugee-site was investigated, two workers independently walked the perimeter and their measurements gave good data on area available – with accuracy more than adequate. Likewise an overall census of likely MP migrants yielded not just body counts, but better data – numbers of healthy bodies for work-parties, numbers of sick and immobilized who would have to be dealt with. And not just numbers but other information - could immobilized people perhaps be carried on the designated foot-paths by small vehicles mostly concerned with food? And if not, then how many litter-bearers would be needed? Knowledge and experience suggested minimum widths for paths, and that set limits to a path's ability to carry foot-traffic. Where to get weather and flooding warnings most reliably and quickly? How to pass the word when a migration became necessary – but without initiating a stampede. Security needs. Those international agencies were identified that might be tapped for expertise and necessaries: the ones dealing with emergency-relief food supplies knew well what tonnage of rice was needed per week for half a million people; how much water and salt to cook it; where to get large-scale cooking equipment; how much secured storage space would be required to hold the materials "ready" but not deployed. All that knowledge and a great deal more was located and tapped. Amazingly, careful approaches to the Air Force eventually got for MP the requested access to unused facilities – a lucky dividend of the country's long-ago opting for minimalist military.

{T_0 - 180 days}
THE MP PLAN DEVELOPS

Progress was assessed weekly during the RedCap meetings with Ramji, Hamid and Lakshmi – the two teachers had thrown themselves into MP with an almost religious fervor. The trio of teachers was quickly nicknamed "Triad" and usually referred to as a single entity.

Making progress could be painful, but a workable general technique was developed quickly. First, Triad would choose two interconnected sub-problems, say shelter and food. Triad plus the two RedCaps would work up a succinct message and minimal request, decide whom to approach first on one of the two problems, work out a presentation that attracted the initial target organization by always proposing a solution to MP's specific problem which solution would also simultaneously fix some other problem for the target itself (usually at no financial cost whatever). Hamid's and Laxmi's professorships were a marvelous help in opening doors. The objective was always to get for the sub-problem a tentative "Yes but... certainly, if you can..." Later came the pretense that such a response was in fact a firm agreement, and using it to gently leverage something slightly more tangible and certain out of the people (or agency) most related to the <u>other</u> "linked' sub-problem. Back and forth, almost like children wheedling favors out of their parents.

They followed a routine – for each sub-problem always a very-well-rehearsed RedCap would be the actual presenter to the necessary meetings and councils. Triad would attend with the RedCap, set things up by giving an introduction, explaining the overall problem to be addressed. The RedCap would then explain the problem in his/her own way and in their own words. The RedCap would clearly present the specific needs of concern in THIS meeting, and outline (a) how the target's own resources and mission fit perfectly into this need, and (b) how the person or agency could actually itself benefit from helping... intelligent self-interest being a huge motivator towards cooperation.

It was astonishingly, almost ridiculously, effective to have a low-caste person, someone personally affected by the problem, do the presentations and field the inevitable questions. In explaining the problem orally, illiteracy was almost an advantage. And it often seemed as if various officials, when they were finally addressed face to face by the needy, could not resist the urge to actually help instead of becoming bureaucratic stumbling-blocks.

Step by step MP built up the necessary connections and commitments for each sub-problem. And all worked together on detailed plans for actually moving people when the need would arise.

In the early weeks of researching sub-problems and generally getting an organization designed and stood up, Lakshmi made contact with a source of money, an anonymous supporter about whom she could reveal only that the person was female, an extremely well known Bollywood star, and quite significantly wealthy. The benefactress agreed to provide funds as needed for the initial organizational effort. As a veteran of other social-activism campaigns, it was she who suggested (and paid for) having weatherproof paper Nehru caps made, in beige but with a broad red diagonal stripe, to be worn as a public sign of the wearer's active participation in MP, and to show off MP's presence and strength. But only those people who committed in some way to the plan, and who agreed to abide by it, were eligible. Being active in MP very quickly became a matter of personal pride transcending obstacles such as the leftover effects of caste. It took just under half a million caps to satisfy the initial demand: within a few days of first distribution the caps were seen (and understood) for over 100 kilometres up and downstream.

6

ON PARANOIA – GOVERNMENTAL AND OTHER

{T_0 – decades, and up through the present}

At partition, in 1947, the British Raj was divided arbitrarily into two religion-based countries, India (Hindu) and Pakistan (Moslem). Pakistan had East and West halves, separated from one another by more than 2,000 kilometres of northern India and Nepal. This improbable and obviously stupid arrangement was a sort of parting gift from the Brits to the subcontinent. (It was not a unique occurrence... the same arrogant self-serving mentality and group of overlords gave us today's lovely Middle East muddle, at about the same time.)

Unified government of the separated halves proved undoable. There has never been any love lost between India and Pakistan, and military sabre-rattling is continuous across their border. In 1971, after the third official India-Pakistan war, East Pakistan seceded from Pakistan and set itself up as the nation of Bangladesh. Bangladesh even today is still quite a weak nation both militarily and economically, and upon its independence decided (unlike the other two) to concentrate its available resources into non-military tasks. Hence the country has developed only a puny military, which is of no consequence or concern whatever to today's heavily militarized and armed Pakistan 2000 kilometres distant: nor is it of any serious concern to giant and militarily powerful India, abutting Bangladesh to the south.

In many important and powerful parts of all three countries' governments, paranoia is rampant and deep-seated, and has been so for decades (more accurately, centuries). In particular Pakistan and India (i.e., Moslem and Hindu) have *ab initio* deeply distrusted and actively despised one another. To each side, the other is absolutely literally the devil incarnate, and would best be removed from the planet wholesale and instantaneously – if only one could devise a scheme to do the erasure without damaging one's own people. In the lower ranks of the national governments and civil services, and especially so in the countries' militaries, those feelings were and are bizarrely strong, yielding not at all to logic or data (the feelings are, after all, fundamentally religious!). Military personnel and government officials spend their entire lives immersed in a cauldron of belief that the other side is forever pushing its destructive agenda, with a perfectly-obvious goal of destroying the other party. Everyone expects, plans for, and sees, only the worst possible scenarios, regardless of the topic... and utterly without regard for data of any sort. Politically, no one on either side can afford to do anything except regurgitate decades-old venom at the other – to even consider that there might exist an innocent or friendly purpose behind an opponent's action would be political suicide. A truly dangerous stand-off situation, and between nuclear powers!

The wonder, then, is NOT that the two countries have survived three wars, but that there have not been more, and even at the atomic level. On both sides there are a significant number of people - especially in all levels of the military - who are convinced of their own country's invincibility and indestructibility (god of course being openly on OUR side!), and who would like to see the conflict opened up and finished off – and most certainly with nuclear weaponry, after all it is expensive and what else is it FOR?

Some small fraction of the overall paranoia is doubtless justified – plots DO exist, at many levels and scales and likelihoods. Unfortunately, the fact that many plots within plots, much subterfuge and double-dealing, are quite real (although often about trivia) provides enough reality that it makes the bubbling collection of imaginary problems much more believable. But MOST of the mistrust is the product of feverishly over-stimulated imaginations immersed in a dangerous, compartmentalized system inherently subject to positive feedback.

The paranoia runs so deep that even within a single given military service there is a pervasive culture of deep mutual mistrust of one's fellows, a feeling that goes like this - *"I and a few friends [in the Army, Air Force, Navy] are the only ones who REALLY understand what's going on here... and only part of the problem comes from the other side – much of the total problem comes from INSIDE my own service, from the disloyal or ignorant people outside my own little group!"*

A wonderful atmosphere in which to operate with nukes in many hands.

{T_0 – about 100 days}

SOMEWHERE IN INDIA
Army Special Considerations Branch
(a.k.a. "SCB" - a TOP SECRET organization):

Colonel Raj was thoroughly pissed off. Plus irritated and angry. With no escape valve available through which to vent his frustration, either on the job or at home. Altogether a most uncomfortable feeling. Most assuredly bad for the stomach, not to mention the intestines and heart. The coolant he self-prescribed was mango lassi - iced of course, one must be civilized! Always custom-made to a traditional local recipe... plus a goodly dose of his own special intensely-sweet citrusy liqueur smuggled in from the Baltics hence obtainable under the counter only, and that with considerable difficulty. And expense. An indulgence in the guise of medicine, always available –as now- from his office fridge.

Raj had just gotten off the phone after an exquisitely uncomfortable ten minutes of being harangued - completely unjustly, of course - by his boss. Said Boss-Man now owned a silver star, was an honest-to-god General Officer. A trivial man unaccountably promoted beyond Raj, but as yet a general of one star only, merely a small difference between him and Raj, one which Raj was working mightily behind the scenes to eliminate. Both men were in the Indian Army's SCB (a.k.a. "Special Considerations Branch") ... a not-well-known organization. No, it was a good deal more inconspicuous than that, actually, deeply secret. Well beyond merely TOP SECRET. The Brits would have categorized it something idiotic, such as "Top Most Hush-Hush Secret". SCB specialized in both making and implementing decisions about socially-unpleasant situations or conditions... hopefully before those problems reached crisis proportions and became uncomfortably visible to the public. The SCB had over time become very powerful, and had today an enormous, and very well-concealed, ability to steer both the national Army, and civilian policy, in directions chosen - sometimes not quite wisely – either by the Army or by SCB itself. On occasion, a specifically military problem would be steered SCB's way for analysis and resolution and planning. But most SCB work was on civilian social and economic problems. Over several decades of working together a fine relationship and strong mutual trust - free of most paranoias - had developed amongst civilian government, military, and SCB.

It was an unusual "potential crisis" that had abruptly put everyone's bowels in a twist, on a minor problem Raj had foreseen and begun work on quite some years ago. Raj had recently fired off a detailed memo to all upper-level bosses, about the people out on the Delta, the politics and economics of that swarm, and the political and military ramifications to be expected when one factored into the equation both global warming and rising sea level. Very large numbers of people out there were coming together – ostensibly to coordinate escape from floods, but who knew for sure what their ACTUAL design might be!? After all, that many people in one place could turn into something politically dangerous or damaging in a blink. Nobody in SCB wanted to hear about it – almost everyone thought the problem trivial, a classical straw-man. Hence Raj's alarm-laden memo had gathered him no friends.

The problem with ignoring Raj's memo was that Raj had done a lot of well-known, well-acknowledged fine work, and had never come a cropper on a problem. Hard to simply round-file a memo from such an accomplished agent. What if he were CORRECT? Raj had made his reputation, which was quite good even if not by consensus spectacular, on genuinely difficult problems, but always problems with no 'flash' - the lack of which is what Raj felt was the reason for his personal lack of a star... which bauble he coveted intensely.

As a 'for-instance' problem of huge importance and almost no visibility on which he had labored mightily (and very successfully!) one might choose the government's need for a variable-intensity, variably-postured stance on (human) birth control. No one-fits-all approach could be even considered, given India's diversity in every aspect of human existence.

Any serious thought about the problems of population control for India immediately foundered on the social costs and ramifications, which proved staggering, once SCB dug into things deeply. For example, if birth control were to become freely available and used without stigma, how was India then to keep up its rapid economic growth and meet that growth's annual needs for lakhs, or more likely crores, of new entry-level, minimum-wage workers? The baby-making machinery MUST be kept well-oiled and up to full speed!

Governmental programs professing to offer birth control for, and controlled by, women were therefore carefully crafted -with SCB's considerable help- so as to always (a) appear to be working well, thereby enhancing both the projects' officials' and the country's international prestige for being so forward-looking, and (b) be actually inconsequential regardless of how stridently advertised and pursued.

Throughout that project, his analyses and suggestions had been very favorably received, a few even implemented unmodified. Raj was certain his own importance and contributions had greatly surpassed those of his one-star boss/rival/nemesis. Yet Boss-Man got the star. The gods moved in altogether unpredictable ways.

The harangue was an accidental, indirect result of Raj having developed, some years ago, a reasonably close friendship, entirely outside of work, with Dr Hamid, the first honest-to-god PhD scientist Raj had ever met. From that acquaintance had sprung Raj's interest in global warming, sea level and the like – including a passing, very cursory interest in the MSL People. All of which interests he had pursued on his own initiative for some years. And things out there were now changing fast. Hence the recent MSL memo. The harangue, Raj felt, was entirely unjust, for Boss Man was suddenly in need of MORE information on MSL People, and, unreasonably, he now chose to be suddenly angry that Raj (un-tasked and unsupported in the effort) could not produce it instantly.

Dr Hamid was a physical oceanographer, with a doctorate from a world-class university in the USA, one of perhaps a half-dozen such people in the subcontinent. A Hindu with the Muslim given-name 'Hamid". Ethnically confusing indeed. From a caste well above Raj's own, Hamid was an expert on physical processes in the ocean - currents, tides, winds, mixing: he was also a well-informed, well-regarded and active member of the international scientific community working on global warming. Hamid was one of the small number of scientists on the subcontinent who could intelligently deal on global scales with the complexities of the interactions of atmosphere with oceans. Studying how wind moves water and how sunshine warms the water and how heat in the water in turn drives the winds, a circular process with no beginning and no end - very Hindu at heart, so thought both Hamid and Colonel Raj.

Whenever possible, Raj, a man vain but hardly stupid, lapped up information from Hamid like a thirsty dog at a dripping faucet. And that scavenged information he dispensed at his job as seemed useful – always given gratis and with a slight aura of mystery, perhaps a whiff of superiority, at meetings both social and professional. ("Where the hell did you find out THAT?" was a frequent reaction by his bosses to his bits of wisdom, a huge ego-stroke.) All of which gave him occasionally an enormous leg-up. On things having to do with the ocean or global warming he gradually became the in-house "go-to" authority.

Unfortunately he was neither subtle nor very discreet: and he was totally blind to the obviousness of his often clumsy attempts to scale various often-slippery social, political and military ladders. Although no-one would ever tell him so, this part of Raj's behavior had been noted and discussed by men each owning two, three, even occasionally four stars… and it was always that behavior, not lack of talent or hard work, which was the reason for Raj's lack of generalship.

Despite Raj's lack of ladder-climbing skills, he was well thought of and liked, and he circulated well, making his presence felt in various ways. Being without the needed formal scientific training, his ability to really understand and effectively use what Hamid told him was extremely limited. And of course, Hamid knew nothing of his friend's need for and use of what he was imparting. Moreover, despite the potential for crossover, that is, of Hamid unknowingly bringing together Ramji and the Colonel, it never did happen, not even in Hamid's imaginings.

Raj decided to handle the developing 'mysterious' "MP situation" entirely on his own: because his memo had been largely ignored, that would be the best way to ensure that any success was entered in his –and ONLY HIS!- 'promotion dossier'. He really wanted that star, and would push hard on the system to make it happen… no shared credit this time! He had a feeling that however trivial others might think this situation, it was likely to go critical and perhaps much sooner than most would guess.

7

PANIC!
WHEN WITHOUT EXPLANATION
14 cm BECOMES 40 cm

{T_0 - 5 days}

Early one morning Ramji noticed a government survey team's big white boat before almost anyone else did – simply because he was up so early to head for school. Curious, he spent a couple of minutes following it, saw it land on one of the semi-permanent consolidated sandbars. A work party got out, carrying odd-looking equipment. He couldn't stay, continued on his way to school. The boat was gone when he passed the spot enroute home at day's end.

That evening, the very first person in line for help asked, quite excitedly, if Ramji knew what the big white boat had done? The boat's crew had worked some sort of magic with their equipment, then they had left, but only after installing a large, tough, rather permanent-looking white plastic stake – it was over a metre tall, with writing on it. The bottom end of the stake flared out and was an anchor – the hole they dug for the stake's anchor was a meter wide and deep, now refilled with the stake protruding. The stake was not going anywhere soon! The small crowd of spectators had been warned by the boat's captain not to touch the stake under pain of a year or more in gaol. Could Ramji please come check this thing out?

It was a government geodesic survey mark: a dimpled black dot on the top of the stake had an impossibly precise latitude and longitude inscribed beside it. And down low, just above the sand, a strong black horizontal mark labeled "Mean Sea Level".

Ramji explained to the crowd that the big stake truly was important and not to be messed with, that it enabled people to figure out exactly where they were on the planet, and most importantly, the horizontal mark told them what the overall level of the sea was. He agreed to do a little demonstration several times over the next couple of weeks, so that everyone could see, no rushing and pushing please. He would bring a graphical tide table (which he would explain to all who attended) and his digital watch (a gift from Dr Hamid, who had been given a much nicer and more modern one by his wife), and they could all observe how the sea rose to a specific height at a specific time, precisely according to the tables. It was also easy to see from the charts the time of day when the water level should be exactly at MSL. That all would depend, of course, on there being no big storms anywhere nearby to confuse the readings, but weather had been fine and stable for some days.

The first two demos went nicely – it took repetition, several changes in approach, and time, but eventually everyone seemed to understand what the graphs in the tide table meant – graphs being so much more efficient and readily interpretable than are tables of numbers (particularly for an illiterate audience!). The apogee of each day's performance was the group interpreting the graphs and predicting the time at which the water level would reach the MSL mark.

$\{T_0 + 1.5 \text{ to } 2 \text{ days}\}$

AFTER THE GIC'S COLLAPSE

Not immediately known to Ramji, on day two of the demos the Greenland Ice Cap partially disintegrated and dumped into the sea a whole two percent of its six million cubic kilometres of water and ice. Two percent sounded like no earth-shattering thing – it amounted to fourteen centimetres of water, averaged over the world ocean, an incremental change in MSL which was in and of itself neither spectacular nor obviously important.

The GIC material entered the sea essentially as a point-source: it took about two days for the water to rise to its new "MSL" around the globe. To raise sea-level, the freshwater from Jayhawk's new digs didn't have to physically flow around the world – rather, the water everywhere would rise the appropriate amount in hydraulic response to the GIC's input. Very like adding a teacup of water to one end of a half-full bathtub, the water level around the world quickly rose to its new level, without the inconvenience of much actual horizontal movement of the added material.

The partial collapse of the Cap was, of course, reported in all media worldwide: Lakshmi arrived at school with maps, newspaper stories, and a short, classically-good BBC special-topic instant video. Ramji listened carefully to the BBC's explanation, picked up on the numbers, and immediately did the calculation – fourteen cm of rise was correct if the new Greenland topographic data in the restructured icecap were good. When he got home that afternoon, he would do his next demo as scheduled, but he would explain this event to today's group, and predict a permanent 14 cm rise in MSL. The showman and teacher in him hoped that the MSL adjustment would have arrived by demo-time, but he had no idea how long the world ocean might take to come to its new-equilibrium level.

Also quite unknown to both Ramji and Hamid was the bottom topography at the outer edge of the Delta, which (as arranged by the Gods for their little game) unfortunately turned the arrival of a fourteen-cm permanent rise into well over 40 cm locally – a one-time-only event related to surges and seiches. The difference between 14 and 40, to someone standing on a riverbank at MSL, with feet wet, is striking... being over knee-deep when one expects at worst a hand's-breadth is, in fact, understandably disquieting, if not simply terrifying. Particularly so if one has only small knowledge of how the overall system works.

Factors other than ignorance often augment peoples' reactions: the Delta MSL population had for some days been hearing rather twisted news of a severe storm supposedly enroute towards the Delta – with a predicted storm surge of nearly three metres, cause for immediate anticipatory migration to the Uplands. The population had no idea that the information on the storm's course and landfall was simply wrong – the storm was certainly strong, but would come ashore so far to the south as to be harmless to the MSL community.

The time of day when the tide would put the ocean's surface exactly at MSL would not vary due to the extra water from the Cap's collapse, because that time was set by the gravitational interactions of Earth, sun and moon. What **would** change would be the amplitude, the height, of the tide at that moment.

Per both Ramji's calculations and the news stories, the people expected the water surface at "MSL time" to be fourteen cm higher than before.

The prediction was clearly badly inaccurate – at the time when the water's surface should be just reaching the MSL mark, it had already gone well past that line, and at precisely the "MSL moment" the water was slightly over 40 cm deep. The more astute in the audience understood that something was out of kilter - but couldn't figure out for themselves just what was happening. They appealed to Ramji, who was equally puzzled... he would certainly think about this oddity overnight. Tomorrow Dr Hamid would be at school, so Ramji could 'squeeze' him for an explanation, which surely the Doctor could provide and which Ramji would bring home tomorrow evening. Ramji's lack of a ready explanation was a rarity and set the audience's fears going much more than Ramji understood – and those fears spread like wildfire.

At school next day, Ramji discussed the situation with Hamid. Hamid immediately suspected bottom topography as the culprit, explained to Ramji the factors and processes involved. Reassured him that these days of course there is uncertainty in much oceanographic data (including the data on the Ice Cap) but that it was unlikely their understanding was off by a factor of two or three. Not to worry, the "14" was a good number, the observed "40" was certainly real, but equally certainly merely a temporary "local" aberration.

When Ramji described his audience's nervous perception of the 40, Hamid and Lakshmi conferred, made an offer Ramji could not refuse - perhaps it would be helpful if she and Hamid were to come over to the village early tomorrow – a Saturday hence no school – and help with the explanation?

Ramji was delighted, would send the Three Brothers with their big boat, to be at the school at six A.M..

Context is always important. The regional weather predictions, and especially the reporting of them, were dismal. The mis-modeled and mis-reported storm was supposedly headed directly at the Delta and seemed certain to bring a two to three metre storm surge, which would require mass migration to the Uplands. In reality, the Delta might get a centimetre or two of excess rain, but most certainly not a three-metre storm surge. It had been quite some time – several years - since a "three-metre" storm. The MSL folk knew the overall MP plan was nowhere near complete, although migration routes had been surveyed and marked. Intense worry about the looming storm, and the abrupt and unexplained 40cm rise, combined to raise widespread panic: uncontrolled migration began the evening of Ramji's discussions with his teachers. Ramji arrived home in the dark, passed the word about tomorrow's visiting teachers as best he could to his own village. He didn't know how widespread the panic had become – nobody knew. Between bad weather-predictions and misunderstanding of the 40-cm rise, panic ensued. Four hundred thousand not-well-organized people packed up and started to move towards higher ground on their own initiative.

At eight next morning, Lakshmi and Hamid arrived – Ramji was just beginning to be fully aware of the impromptu –and unnecessary- migration now well under way. Triad spent most of the day collecting what bits of information they could, and pondering what might be done to reverse the unnecessary and probably dangerous migration. Unfortunately, corrective information is harder to distribute and almost always less impactful than the original bad news, however distorted the news might have been. Attempts to calm the peoples' fears about the weather foundered on the very poor media presence on the Delta – almost no radio receivers, no such thing as cell phones.

Triad could do nothing that day except ready themselves for tomorrow. The plan was to next morning gather Ramji's own entire village of many hundreds around the big-crowd podium out on the flats, explain to them the weather reports and the anomalous 40 cm, and send them off as calming emissaries to various MP routes, hoping to turn back the human tide. King Canute's long-ago attempt to reverse the salt-water tide had been equally successful. At least, in this the first MSL/MP exodus nobody got hurt, to Triad's great relief.

This unplanned mass migration was spotted and poorly documented by Bangladeshi photo-recon drones doing their clandestine over-flights of the Indian mudflats, the data being fed, unauthorized, into Bangladesh's thoroughly paranoid intelligence services for analysis. And thence, also unauthorized, to Pakistani intelligence.

8

"INTELLIGENCE" MACHINATIONS: INDIA, BANGLADESH, PAKISTAN

{T_0 – Some weeks before migration}

(a) IN INDIA: It had been the spectacular flood of red-striped Nehru-caps during the test-assemblages and impromptu rehearsal that finally really brought MP to the center of Colonel Raj's attention. The Delta rabble was apparently NOT rabble, and seemed to be developing a uniform – the 'rabble' must therefore be going military. A classic situation rapidly a-building – first came some organization of the movements of large numbers of people (perhaps disguised troops?). Organized movements implying central control! Large movements of people were always suspect... humans *en masse* were particularly susceptible to demagoguery and control, thence directable into actions leading to great social mischief. Then along comes the proliferation of striped caps – which were obviously their uniform. What next?

He was concerned because recently he had (again on his own initiative) shifted much of his attention to the social instabilities caused by sudden movements of large numbers of people in the Delta, as a result of changes in river levels. That redirection of interest towards MP was greatly dependent upon his source of data on global warming, tides, and sea-levels, namely his friend Hamid. The Colonel's personal paranoia made him wonder what was going on amongst the MSL People. Using his rank and position, Raj arranged for a little spying, but it had been ineffective. Forcedly-casual reconnaissance from patrol boats yielded little of help, other than the fact that almost 100% of the population visible along the banks was wearing those cockamamie red-striped caps. By calling in some personal chits, Raj managed to schedule unauthorized and undocumented over-flights by low-tech photo-recon drones. Those flights produced the same result as did the boat-surveys, but gave it a more areal dimension: the big mudflat was TEEMING with those damned caps! And EVERY where... not just at the water's-edges where the patrol-boats had seen them.

A movement was afoot, obviously – the gods help him! That made the Colonel even more nervous – he really did not want a "movement" to occur in his territory and on his watch. That way, he knew, lay error, mistakes, demotion and disgrace. One could actually -horrors!- lose one's pension in dealing with a "movement". Raj had himself seen two such cases.

Genuinely nervous now, and fretting over the lack of good intelligence, Raj considered the idea of on-the-ground infiltration, real spies (against his own citizenry – but that was hardly a new situation in intelligence!). Infiltration, it turned out, would not work: the MSL People were infinitely suspicious of strangers, and had a largely closed society with its own ways of doing things, things from properly eating a herring to subtleties of language. Local "ways of doing" utterly unknown to the outside world including Raj's intelligence resources. There was simply no chance of slipping agents into the population inconspicuously enough to collect useful data.

Colonel Raj was constitutionally incapable of seeing nothing where there was in fact nothing. He and his compadres MUST see something, lest their superiors think them incompetent, or at least insufficiently occupied. Had he not been deeply wrapped in the arms of paranoia Col Raj could, of course, have found out in minutes who Ramdaas was, and thence gone instantly from him to Hamid and Lakshmi, and onwards to the MP itself. Which (had he done so) reasonably should have allayed his fears, the intent of a population to avoid floods hardly being the same as fomenting some form of insurrection. On the other hand, knowing further that Hamid was part of this movement could have, but almost certainly would not have, cooled his bubbling paranoia. He would have seen not an innocent connection and a social problem, but rather something quite sinister.

And sinister Raj already had in abundance - an apparently rapidly-building, low-caste mass-movement impenetrable to his efforts, perhaps going militaristic, with unknown specific goals, rumored to have a very young and charismatic leader. All this on Raj's own chosen turf. Not a good thing. Coils within coils - when Raj sent to Boss-Man the information from "officially-nonexistent" over-flights, that material was instantly forwarded by Boss-Man to his own counterpart in Bangladesh, of necessity a party interested in any Indian goings-on out on the Delta. The attitude was that mutual help, utterly illegal yet universal, could only benefit both parties.

(b) IN BANGLADESH: Bangladesh had its own perpetual, self-renewing governmental and military paranoia, although it was definitely an order of magnitude less concentrated than the Indian and Pakistani versions. Bangladeshi military and other governmental agencies took note of MP's activities, and were worried – after all, the country was already the repeated recipient of half a million temporary INDIAN residents for whom no proper preparations had ever been put in place by either nation. Bangladesh also had an intelligence problem, although not so severe as India's – in fact, the Bangladeshi intelligence community had much the better informal connections to MP than did the collectivity of Indian authorities. Bangladesh depended for its information on occasional over-flights of the Delta by ancient drones carrying simple video cameras. The resulting imagery was low resolution and covered only a tiny fraction of the Delta surface... but it could show people and, with the drones down near the deck, details such as the sudden proliferation of striped caps.

(c) IN PAKISTAN: Some two thousand kilometres to the west, Pakistani interest in any Bangladeshi problems was insignificant, and Pakistan had little reason to develop its own intelligence about the far-away Delta and its people. Most of the Pakistani information came laterally, being informally shuttled through the paranoiacs in Bangladesh – straight into the hands of the lowest-level Pakistani paranoiacs.

Thus (at a relatively low level) both India and Bangladesh were aware of and concerned about MP. Paranoia whispered into Bangladesh's ear, "Such a mass-movement in such a place simply MUST have governmental support!" This despite centuries of similar migrations. Bangladesh's paranoia took the form of wondering what nefarious plot the Indians were hatching within this present migration of half a million Indians into Bangladesh yet again – dammit, at the very least, Bangladesh couldn't go on forever taking care of that flood of Indian-citizen refugees – it was annoyingly expensive, disruptive, and unfriendly on India's part not to take care of their own damned citizenry. Besides, perhaps this time it might be a genuine 'peasant's rebellion', aiming to take Uplander territory permanently!

9

GONE MISSING:
ONE (1) ATOMIC BOMB
MODEL N-17, 200 kilotonnes

{T_0 - three months}

In those human affairs not driven by logic, fear and paranoia usually rule. The smallest, most apparently innocuous occurrences can have huge consequences... especially when they invoke the laws of unintended consequences. Which laws are also known as the Gods' game-rules. One Friday afternoon on a large Pakistani airbase near Lahore, a supply sergeant, hurrying to close out his workweek, left a classified document on his desk... not a major failure of security, so it would seem, for it was merely a catalog. A reasonably thick catalog, of some of the more esoteric armaments available through the military's supply system, including both parts breakdowns and whole systems. It also included the military's stock numbers for whole systems and for all those systems' component bits and pieces.

Late that evening two very young electronics technicians entered the building, to which they had a key because they maintained equipment 24/7. They were seeking parts with which to repair a radio. Having quickly gotten what they needed, they wandered the unmanned building for a while, just snooping. And found the catalog. Which of course they browsed eagerly. To their utter amazement, under "ordnance" they found listed several varieties of nuclear weaponry – listed exactly as if those nukes were of the same importance as a vacuum tube or a bunk-bed mattress. The men had been thoroughly bored all day, anything for a brief diversion. They had a giggle over the listings. Then one said to the other, "Hey there, Private First Class – you and I know how to order stuff, we do it all the time, just write out a chit with an item's stock number on it and drop it in the Sergeant's tray." Then, scanning about, he went on – "Hey! Look - there's a pad of requisition forms right here! Let's tweak the Sergeant… read me the stock number for the fanciest nuke in the list." Laughing, they ordered one N-17 nuclear bomb, signed an illegible left-handed signature on the form, put it half-way down in the pile in the in-tray.

They left in a howl, envisioning the Sergeant's face when he got to the requisition.

The Sergeant was busy next Monday morning, and simply rubber-stamped and forwarded the bundle of weekend requisitions without bothering to check them. The techs were disappointed at the silence from their Sergeant, but chose not to enquire, sensing potential trouble down that path.

The Pakistani Air Force's security measures for such weaponry were astoundingly lax: when a properly signed, stamped requisition arrived on yet another Sergeant's desk, aboard a different airbase far from Lahore, he studied it briefly, found it in order, nothing weird – after all, where else would one send an aerial nuclear bomb save to an airbase? His duty, beyond which it was always dangerous to step, was specifically to fill properly-submitted orders for materiel – he was not an order-checker, although if the destination had been a ground-pounding army unit, flags might have gone up.

No flags. The second Sergeant okayed the requisition – again literally by rubber-stamp - and shipped the bomb from its storage area … its personal storage space was originally for truck tires, under a tarp, in a half-garage, along with standard ordnance, mostly 500-pound simple iron bombs.

Shipment was by panel truck with driver and one guard, said guard being all of nineteen years old and without knowledge of their cargo. He was armed with an ancient .38 caliber revolver loaded with surplus Vietnam-era ammunition. The bomb was fully crated: the only ID on the crate was the stock number and a stapled shipping label, nothing describing the device inside. And certainly no safety warnings! It was dropped off at the two technicians' electronics supply building, where it was signed for by a random person. There, it being obviously not a standard electronics part, the box got shoved into a corner, unopened and unnoticed. Where the next day it was discovered by the original incredulous pair of techs. Who thought the whole situation so hilarious that they got drunk that evening (alcohol being both officially prohibited and religiously prohibited, yet universally available in Muslim Pakistan). The two techs, drunk, mentioned the caper to a couple of friends. Next day, that story trickled upwards and arrived -amazingly- in the ears of a four-star general of singularly ill repute – his nickname, occasionally used to his face, was "Daisycutter", and thought to be well deserved. The real-life 'daisycutter' is a particularly evil form of shaped-charge antipersonnel land-mine – when triggered, it first hops into the air to knee-height, then explodes in a horizontal plane – the entire intent being to remove human legs, and preferably NOT kill the amputee – much better the wounded man should survive and require the services of four or six comrades, all of them thus being effectively removed from combat for some considerable time.

Even in the service, Daisycutter was notorious for being paranoid; he was also a well-known outspoken advocate of extremely strong responses – military responses – to situations arising. Diplomacy, in his view, might be reasonably indulged in, but only AFTER bombing had commenced. Daisycutter quickly traced the N-17 story back to the two techs, who reported to his private office on the double. There, they were sworn to secrecy; threatened [utterly believably] with an immediate firing squad if the secret leaked; and promoted one pay-grade each, retroactive two months.

The bomb was picked up that afternoon by Daisycutter's troops, and it effectively disappeared – given the near-lack of both security and inventory control, the bomb's absence was never noted until too late. Daisycutter chortled to himself over his good fortune – he was quite certain that he could find a good, patriotic and rational use for such a toy. There was no hurry to decide exactly what to do. Patience would surely be rewarded.

A SHORT PRECIS TO DATE

Although the MP is still a nascent 'work in progress', MSL People have started a migration, mistakenly believing in the wildly-incorrectly-forecast major storm and being quite certain that there is a huge change in sea level imminent, to follow on the heels of the unexplained 40 cm rise in MSL. Using its woefully inadequate sources, Bangladeshi intelligence does notice the migration and expands its originally spotty and casual drone coverage – so that Bangladeshi drones are now openly overflying Indian territory, which is loaded with Indian citizenry on the move towards Bangladesh's higher ground. Indian long- and mid-range radar is incapable of spotting small low-altitude drones at distances of the width of the Delta – even if they were looking for them, which India has no reason to do.

THE TWIST

There exists a secret mutual defense treaty between Bangladesh and Pakistan – 'secret', yes, at least so labeled, but known to many senior officers including Daisycutter. Pakistan –so says that treaty– will come to Bangladesh's aid if and when Bangladesh has a problem with India - with the exact nature of both any triggering 'problem', and of any response, being left open. The Pakistanis' information on mass movement of MSL People comes from both Bangladesh and India by sometimes devious routes, but goes only to the lowest and most paranoid level of Pakistan's military, where it is interpreted (as usual, by paranoiacs). Among those interpreters a belief instantly arises that the mass movement is armed, organized militarily, and enroute to invade and take over Uplander territory in Bangladesh. The interpretation is dire, foresees a big invasion, a major international incident. All of which is instantly agreed with by Daisycutter, the tentacles of whose 'personal intelligence' reach across all levels and draw paranoia from them all.

Due to paranoia in thick layers, that analysis does not go UP the chain of governance, but rather sideways to the equally paranoid lowest levels of the Pakistani military: Pakistan is as yet totally uninvolved, but there the analysis is received as the gospel truth. Then there issues to Pakistan from Bangladesh's intelligence and military basements an unofficial but unsubtle invocation of the secret mutual defense treaty: someone in Bangladesh suggests to someone in Pakistan that a military response by Pakistan might be in order, since unarmed Bangladesh is incapable of it herself. Once again that suggestion percolates through, and in fact comes from, the paranoia in the various basements, but as seems usual does not trickle upwards to the appropriate levels of government or military, seeking a response. Instead, the appeal goes sideways, lower-level to lower-level, and by dumb luck and awkwardness it also gets to Daisycutter, who has his hands on an N-17 nuke and who thinks that diddling India is an utterly FINE thing to do.

Daisycutter decides to goad all the upper slowpokes and "fraidy-cats" into action. In total secrecy he designs, assembles, and okays The Mission.

10

"THE MISSION": GENERAL DAISYCUTTER, COLONEL HAKEEM, & NUKES on the MUDFLATS

{T_0 + about 1.5 to 2 days}

A LOOSE CANNON in the NUCLEAR AGE

As explained earlier, and here re-emphasized because it is the underlying cause of The Mission and its consequences, all levels of government and military, in all three countries, are forever rife with paranoia – about the other countries and their possible plans, certainly, but also about internal politics and power struggles. The lower levels of the military are the most volatile. Their manifold and rapidly-changing paranoias feed on one another, growing like aggressive, metastasizing cancers. Conspiracy theories abound.

Initially, someone in Bangladesh's trivial air force had seen the country's poor-quality aerial-drone photos of the concentration of active MSL People in the Delta, and had instantly assumed that they had been intentionally massed there by India. Massed, of course, with evil intent – and while that intent was unspecified, it most likely included a land-grab. It made no difference to those men's reasoning when they were actually shown extensive brand-new photo coverage (by their own Bangladeshi drones) which showed that those masses (a) were composed entirely of civilians who had no weaponry, and (b) who likewise had neither roads nor any buildings taller than two stories, all of their very few structures being temporary - made of lashed bamboo.

Bangladesh's wildly over-amped interpretation of those slender data was that they indicated an India-inspired mass civilian assault into Bangladesh territory.

Some of the analysts knew that at least SOME of the data were coming from India (via Col Raj) – and chose not to raise the question, "Why would India be both sending us information about the movement, and also orchestrating the movement?" The incongruity was held to be proof of Indian perfidious intent – and the level of paranoia ratcheted up several clicks.

At any rate, this conclusion of "actions by India, with bad intent" was signed off upon by several layers of intelligence, then passed not upwards through the normal Bangladeshi hierarchy, but rather laterally, to equally low-level, equally paranoid folks in the Pakistani Air Force. There, the story and data-misinterpretation were further embellished before the materials percolated upwards. Straight to Daisycutter's ears.

With the data and analyses came an unsubtle reminder (somewhere between panic and hysteria) of Pakistan's mutual-defense treaty obligations to Bangladesh, and a suggestion (not quite a request) that some military intervention by Pakistan on Bangladesh's behalf would be very much in order, thank you very much, soon is good and sooner is better.

It was of course completely unclear what sort of intervention might be appropriate, and exactly when, where, and on what scale.

Daisycutter thereupon decided, entirely on his own initiative, to implement his personal style of preliminary negotiations, thereby cutting to the chase and circumventing all the top-level 'fraidy-cat' political generals and national civilian political leaders, who could never get up the guts to do anything useful or definitive.

THE MISSION

The Mission was designed and authorized in closest security by Daisycutter, then quickly assembled. When he asked for recommendations for a pilot to assign to a super-secret mission, all votes went to a certain very deeply trusted senior pilot, Col Hakeem.

Shortly, one lonesome Pakistani Air Force F-16 (General Dynamics' finest product) took off from Lahore in NE Pakistan, carrying every external fuel tank that could be fitted. It was piloted by Colonel Hakeem, who had nearly 5000 hours in type. Only two F-16 pilots in the PAF had more time than he, and those two were test pilots who flew daily, sometimes several times in a day.

Hakeem held almost every imaginable security clearance, and had flown quite a few oddball, highly secret missions – to the point that doing so was almost his specialty. Hence his universal recommendation. Today's event seemed to be a very special mission, super hush – he hadn't yet been told the nature of or rationale for the mission: most likely with this much secrecy involved, he would never find out. Nothing unusual in that. He knew that he wouldn't even be given the geographical coordinates of his destination until he'd refueled 2500 kilometres from Lahore, in the tiny area reserved for military craft, over in one far corner of Bangladesh's Dakar airport. In fact, it wasn't clear whether he should be thinking about "destination" rather than "target" – one could rationalize either term. There were no externally-mounted antiaircraft missiles and no Vulcan cannon gun-pod, he noted, so presumably the powers that be didn't expect him to engage unfriendly aircraft. That was a good thing.

He was reasonably certain, however, that he'd been selected for some pivotal (although undisclosed) role in the ongoing and ever-escalating histrionics between India and Pakistan – and perhaps also for a role in the less strident but always-present tensions between India and Bangladesh – else why was he being allowed to refuel in Dakar? The facts said something, but unclearly. To begin with, all identifying markings on his plane had been painted over – national insignia, tail number, nose-art, squadron ID – all gone. He snorted in disgust, damned the idiots upstairs. The extra drag due to the roughness of that sloppy paint would take almost 25 knots off his maximum speed. Not to mention what the paint's chemistry would do to the F-16's stealth capabilities. IDIOTS! Also, he had been ordered to take a screwball flight path - over 1400 kilometres of it was actually through the Himalayas and over a considerable expanse of not-Pakistani territory, flying on the deck, using his ground-avoidance radar. He would operate under strict radio silence. Apparently he was not to be seen (or heard, or identified), particularly by Indian radar. Second, he was pre-cleared to land and refuel at the Bangladeshi capital's airport, which suggested strong complicity between Bangladesh and Pakistan – but on exactly what, he couldn't venture a guess. The only thing even loosely binding the two nations was a fanatical distrust of India.

So, all things considered, Hakeem's mission for today -although odd- did not smack to him of anything more profound. Maximum drop-tanks for extra fuel had correctly suggested a long flight, although such tankage had so far been overkill - he'd not gotten even close to dry. The only anomalous, worrisome, or scary thing about the mission was the single piece of externally-mounted weaponry his fighter carried... an advanced (Model N-17) atomic bomb. Hakeem knew to a certainty what that sharply-pointed metal cylinder was: he and his colleagues had practiced for years with replicas – always those practices had used bombs painted all-over blue, the world-wide signal for "inert practice munition, no explosives". The converse message, "LIVE MUNITION", is always proclaimed by a screamingly-obvious yellow circle around the nose of every live device, from bazooka rocket to 500 or 1000 pound general purpose bombs, and all the way up, presumably, to the the biggest atomic and even hydrogen bombs.

The presence of the bomb surely signified something- but what? Perhaps he was literally a messenger service, a package delivery man in a fancy vehicle? Why some idiot in Pakistan's government or the PAF might want to hand over such a device to the Bangladeshis (or to anyone else!) he could not fathom. But of course such fathoming was NOT part of his duties – more than that, it was actively to be avoided.

Today's bomb on his F-16's external hard-point was unpainted shiny stainless-steel, and wore the evil yellow circle. He had contemplated it, of course, during his walk-around pre-flight – but his job was NOT to ask questions. Pakistani military aircraft were often loaded with their mission ordnance, and then launched, with the pilot completely ignorant of the actual task until he'd been vectored in on some target unknown to himself, a target most often selected in a dark room far away by anonymous "trolls" who had never personally successfully flown even a paper airplane.

Indeed, he knew the device well – only seven feet long, double-tapered and a mere 40 cm in diameter at the middle, it was the most modern of Pakistan's atomic bombs. It had a variable yield – 20, 50, or 200 kilotons. The variable feature was nicknamed

"Dial-A-Yield In The Field"

The explosive force was selectable by the pilot, upon instructions from the trolls. The other bomb-parameters over which the pilot had control were the arming switch (on/off), the bomb release button (drop/no-drop), and the altitude for explosion, which was settable in 500 m increments from ground burst to 3000 m – always per instructions from mission control. Plus, of course, the pilot himself, who was the ultimate target-selection-device.

The thought of so much destruction in such a small package had always raised the hair on the back of his neck: it was a major feature of Hakeem's ongoing internal debate over whether he should continue his career as fighter pilot... he really, really did not want to ever have to deliver a live atomic bomb onto a live target. One-on-one fighter combat he understood, every civilization had a warrior caste or tradition. And likewise he understood and countenanced ordinary munitions such as bombs, rockets, gunfire, and even the evil nastiness of napalm ... but the concept of using the mass destruction of civilian human beings as a weapon, or as an instrument of policy, did not compute for him. The perpetual, vitriolic, religion-driven (hence by definition totally irrational) sabre-rattling that accompanied India/Pakistan relationships was always the most worrisome thing in his milieu.

Refueled and ready for takeoff, Col Hakeem punched his ID into the onboard computer: the computer checked the plane's coordinates by GPS, checked the time, and displayed his instructions.

The instructions were of little help in clarifying his task: "Proceed to coordinates below, upon arrival orbit at angels three, after one full 2-degree-per-second, three-minute standard orbit anticlockwise to confirm location, this system will provide further instructions."

Hakeem's stomach clenched – this was unprecedented, unpracticed. What the hell?

He hesitated for a full minute, checking and rechecking his instrument panel. Checking and rechecking his conscience. Finally he shrugged mentally, said to himself "Well, Colonel Hakeem, this is after all your job, so let's at least go see what's at those coordinates – maybe the sneaky damn Indians are doing something wily again and could do with some reciprocal threatening – but it sure does look to me like the coordinates are in the middle of the goddamned Ganges Delta!"

It wasn't fifteen minutes flying to the coordinates. He'd been correct with his guesstimate, the specified location was well out in the Delta. Approaching the given coordinates, he wasn't flying a full kilometre above the mud, and the entire flat, as far as he could see, was swarming, almost seething, with people, all in vaguely coordinated motion towards Bangladesh. Moving in what looked like organized streams using defined pathways. Flying just above stall, he swept over a collection of perhaps several thousand people, packed together into a tight cluster, all squatting: there was a platform in the middle of the cluster, and on it stood three figures.

Hakeem watched the three heads, then their bodies, turn to follow his plane's progress. The platform was at the proper location to serve as pivot-point of his three-minute standard circle: he banked, pointed his left wingtip at the platform, slowed to "ridiculous", and began his required 360.

He studied the crowded scene, wondering just what the three people on the platform were thinking about having his wingtip so obviously pointed right at themselves? It was crystal clear to him what was going on down below – people en masse were fleeing the river. It must be either in flood or about to flood, he thought. Tens of thousands, probably hundreds of thousands, on the march towards, surely, somewhat higher ground. And now every face was turned upwards to stare at Colonel Hakeem in his high-tech flying carpet. Hakeem found it mind-bogglingly disquieting, being the focal point of attention for so many people. His enemies and targets had always been both impersonal and (usually) invisibly distant. Not so just now. His stomach knot tightened. He would complete his orbit in a few seconds.

11

RESPONSIBILITY: THE ULTIMATE MANTLE

Then the fighter's computer system came on.

ORDERS!

He stared, his mind seemed frozen: the display read "Position confirmed. Deliver to this location, immediately, one N-17 set for air burst 1500m, yield 200k. Repeat airburst 1500m, yield 200k. Approach from west to east, standard over-the-shoulder delivery. Post release, exit at max V, lowest altitude. Return home via flightplan Bravo in onboard memory. Squawk mission control 30 minutes notice for aerial refueling if needed. Otherwise, maintain radio silence until within 140 kilometres of home."

Hakeem couldn't believe the message, the instructions. There simply had to be a serious error, perhaps a huge overall error, perhaps merely an error in the targeting coordinates, but ERROR there simply HAD to be. While his brain tried to sort things out, held his course, going into a second orbit. This order was IDIOCY. SHEER LUNACY! Hakeem had no idea what the muckymucks who put this mission together were actually trying to accomplish, but a 200 kiloton low-altitude airburst HERE was unthinkable. For god's sake, there had been not even a HINT of war when he strapped on his plane this morning!

Besides - as far as the eye could see, just people. No structures, no equipment, nothing military. Universally, military doctrine held that big rare bombs and other special weaponry were reserved for the highest-value military targets, which this most definitely was NOT. Just people by the lakhs, perhaps half a crore. No cover. No buildings. No roads. NOTHING but mud and humans. He finished his second orbit, wingtip still pointed squarely at the squatter-surrounded platform. Still three figures on it, a big man in a white dhoti, woman in a scarlet sari, a smaller male wearing a dhoti and a red cap and holding a bullhorn in his hands. All staring open-mouthed up at him... Ramji and his teachers, busy educating the whole of Ramji's village as envoys to go out to try to stop the unnecessary panicky migration.

Then, forever (and justifiably) suspicious of the entire structure in which he was serving, the thought popped up that just maybe he, Col Hakeem, thirty-year veteran of his country's Air Force, might be being set up as the patsy, the sacrificial lamb, for some sort of bizarre grand-scale terrorism. But - by his OWN GOVERNMENT?!

As he began orbit number three, he decided to hell with radio silence – something was so obviously awry that he must call for clarification. No way on god's green earth could he chuck a nuclear bomb at half a million people clad in rags, all of them just trying to avoid a rampaging river. He chose the satellite-relay frequency for headquarters, broke radio silence. That break all by itself, he knew full well, signaled the end of his career. No matter what, this was his last flight in his favorite bird. Oh, well.

"HQ this is Tango One. I am breaking radio silence on my own authority. Come in please."

A long pause, then "Roger Tango One. This is Daisycutter speaking. Have you delivered the N-17?"

'Daisycutter' - the single most feared general in the Air Force, perhaps the most dangerous man in all the country – four stars and monstrously unstable, well known for making, and carrying out, threats of extreme violence in order to get the behavior he wanted from subordinates – and from equals and superiors also. Threats up to and including (said the rumor-mill) murder. Many a clandestine late-night beer-fueled discussion in the PAF had centered on whether/how to get rid of him. But meanwhile, he was The Boss.

Hakeem took a deep breath. "Negative, Daisycutter. There has to be a serious mistake in the coordinates. I am orbiting the designated delivery point at angels three. Delivery point is in the middle of the Ganges Delta. Nothing in sight in all directions except half a crore of people who seem to be fleeing from flood-waters. They are all heading towards the nearest high ground. There is not a single vehicle in view – all of these people are afoot. Out here there are no repeat NO military structures or equipment of any sort, not even a policeman with a sidearm. I need confirmation of the targeting instructions."

He took a deep breath, and fully committed himself: "I also require an explanation for this entire mission. I am now flying an unmarked war-plane in Indian airspace, over Indian territory, apparently without their permission. That is an exceedingly warlike provocation, and I am now a perfectly legitimate target for Indian forces. Does a formal state of war exist between Pakistan and India? It did not several hours ago when I began this mission."

There was a long hiatus, then "Daisycutter here. Your job is to DO, not to THINK. There is no error in the coordinates or in your instructions. The reasons for this mission are not mine to disclose to you, nor is it your job to question my authority and the authority of our collective superiors. You will deliver the N-17 NOW, Colonel Hakeem. Immediately! Is that clear?"

Hakeem took another deep breath, the oxygen whistling through his mask and into his lungs. This was impossible! Some goal, some plan, had been devised and was being implemented by a maniac and now he, himself, was being drawn in.

He took his stance, accepted the responsibility - he could not, and he would not, play this hand. No sane person could!

Thinking very fast, Hakeem spoke – carefully, precisely, slowly... his voice tight with self-control. "General, delivering the N-17 will do nothing except kill half a million humans. Nothing of military significance exists down here. NOTHING. Mass destruction of human life, especially absent a declaration of war, is specifically illegal and that is known to both you and everyone in the room with you – and all of them who are not following my lead are as guilty and complicit as yourself. Your order, General, is in direct disobedience of the Geneva Convention and I am required by international law to refuse to follow that order. The principle of being required to NOT follow a clearly illegal order was established forever at Nuremburg. People have been hung for following illegal orders. Refusing that order is my bounden duty to my country, my profession, and to humanity."

He paused, swallowed, and continued: "Herewith you have my resignation of my commission, effective NOW. I cannot and will not kill half a million people and have my name instantly become the most popular and heartfelt mother curse of mankind's next five thousand years."

Daisycutter went silently berserk internally: his calm outer persona said "Tango One, Colonel Hakeem, this is Daisycutter. Do not presume to explain to ME a soldier's duties under the Convention. There are higher duties. Let me be perfectly frank and clear. This mission has been sanctioned at the highest levels. Neither you nor I can change it in any way. If you refuse to comply, to follow your orders as given, then when you return I guarantee that I will personally meet you at your aircraft with a firing squad – you will not set foot alive on our sacred national soil. No such a traitor could be greeted otherwise. Not only that, Hakeem, but I will see to it that your entire family faces exactly the same treatment. You know my reputation – I mean this!"

Hakeem laughed into his microphone - a huge weight seemed to have been lifted off his shoulders and he spoke freely: "Daisycutter, that threat confirms for me your reputation as the biggest horse's-ass in the entire Pakistani military. You aren't even smart enough to have done your homework – I have no family whatever. Your threats are as empty as your head. If I were there with you I would long since have arrested and confined you. I hope others who are listening to this do so immediately." A pause, a snort, and Hakeem continued: "In fact, if I were there, I would very likely have already shot you myself."

Then, "Something very important for you to note, Daisycutter, is that I now own this airplane, and also everything in or hung upon it… and it is I, ex-Col Hakeem, not you and not whoever dreamed up this business, who will determine just what happens with this lovely equipment. I take that responsibility upon myself. And as to consequences of this mission so far, I suspect some serious international and internal repercussions. Surely, General, you are not so dense as to believe this entire conversation has not been listened to and recorded by Indian personnel? Not to mention the Americans and Russians! I suspect someone other than me could well be facing a firing squad shortly. Care to guess who I think that might be? I might volunteer to be on the squad. And for those standing by and just listening, if you do NOT take Daisycutter into custody, you will be standing beside him in front of the firing squad, believe me! You have all been educated on the Geneva Convention – USE that training!"

When there was no immediate reply, Hakeem shrugged, the motion revealing to him how drenched he was in his own sweat. He thought how Daisycutter had been correct - accidentally - about the existence of higher duties. It occurred to him that being appointed, unasked, to temporary god-hood lay upon himself as a most uncomfortable and ill-fitting garment.

He had, indeed, decided that this plane and its bomb were, as of now, HIS property, his personal responsibility - and nobody else would decide what to do with them. He would choose the course of events, play **Little Tin God of Atomic Bombs** for a few minutes. It was high time to make sure that nobody on any side of any disagreement would ever lay hands on his particular, now-personal pointy-ended stainless-steel bottle with its contained evil genie!

Without further thought, he peeled out of his orbit, advanced throttle to full sans afterburner, and headed east, out over the utterly unpopulated outer third of the Delta, where the mud was soup, not even walkable, where there were and could be no settlements, not even floating fisheries. He steered slightly south to gain distance from the long line of commercial ships entering and leaving the river-mouth's deep channel.

Hakeem jettisoned all drop tanks save one, went to afterburner, accelerated to his maximum speed given the thick atmosphere and the drag of his final drop tank. As he accelerated, he punched the "ARM WEAPONS" button. Bomb and plane compared data: on came a bright red light, blinking "ARMED! ARMED!" Next, he unlocked the 'drop" button, set burst-altitude to maximum, dialed the yield back to minimum. He took two long breaths, rechecked – armed YES, release engaged YES, burst 3000m YES, minimum yield YES. He pulled the nose up and around into a long, slow, near-vertical barrel roll. He waited for the proper moment, knowing that his body would do things properly after so many years of practice. All the necessary needles were holding steady, all properly aligned for this maneuver, which he'd practiced hundreds of times.

Wait.

Wait.

Almost……

NOW!

Passing through 3000m, Hakeem thumbed "DROP" and felt the tiny bump as the N-17 left on its own, headed nearly vertically upwards at almost 1300 kph. It would rise to about 9000m before starting back down towards burst height. He horsed Tango One's nose around and headed for the deck, going southwest. No more than 100 m above the sea, at over Mach 1.4, he was counting seconds – best guess would be about 70 to detonation. At this speed he would be over 25 kilometres from the blast, and most likely wouldn't even feel the shockwave, but the flash would be devastatingly intense. To avoid being blinded, he reached up and dropped his helmet's darkest sun-shield into place, then flipped the rear-view mirror. Such an anachronism, he always thought – a hundred-rupee rear-view mirror, a device likely invented by the second person ever to own a primitive automobile, yet an absolutely critical piece of this 80 million dollar ultra-high-tech machine. And the damned mirror itself could also kill you in more ways than one. David and Goliath, alive and well and still arguing.

At minimum yield and maximum altitude, high above the outermost Delta, the bomb should do the least possible damage. Perhaps, he thought ruefully, this little incident would help the whole world's public and its mostly-idiot representatives to once again reconsider the business of nuclear weapons. The influence of Hiroshima and Nagasaki was clearly being steadily eroded, the lessons moral and political and military were being forgotten, or transmogrified into something ever less effective.

Twenty seconds remaining, he guessed. He climbed to 1000 m, a bit of insurance, some recovery altitude if needed.

Out on the mudflats of the Delta, half a million pairs of eyes had followed Hakeem through his low-altitude orbits, wondering fearfully what was going on. A very few -including Hamid and Ramji standing on the platform- had recognized the aircraft type, but –a sinister fact- the plane sported no national insignia. Those few knowledgeable people were acutely aware that Pakistan flew the F-16, and that neither India nor Bangladesh did so. Therefore, this had to be a Pakistani plane, lack of markings notwithstanding. And THAT made those knowledgeable few particularly nervous, knowing also that the mudflat on which they stood was Indian and not claimed by Pakistan, so why was that country's warplane at least 2000 kilometres from its home, out here in Indian airspace? All heads turned to watch the plane accelerate eastwards, where it quickly disappeared from sight. A huge collective sigh of relief washed over the migrants.

At 72 seconds after release, the N-17 did what it had been built for and told to do.

But Hakeem's bomb killed nobody.

It merely added, quite briefly, a second sun to India's northeastern sky.

Although four or five times the size of Hiroshima's misfortune, this was a remarkably clean nuclear burst – the fireball never came close to touching down, the bomb itself provided the only solid material from which to make fallout... in the grand scheme of things, a trivial quantity. The down-and-out shock-wave found nothing but mud and water to annoy. Tens of kilometres away, a few of the closest ships in the channel lost the larger flat windows on the blast-side of their bridges, but suffered nothing more serious. The direct-line slant range from fireball to humans on the mud was quite a few kilometres and nobody was seriously overexposed to direct radiation. The event made for years, decades, lifetimes of story-telling... after all, what could be more exhilarating or unusual than being atom-bombed... and being MISSED!?

The luckiest MSL People were those who gave up trying to see a now-vanished airplane, and turned their attention back to their path towards Bangladesh and its Uplands... and the lucky included most of the migrants. Only a few of those who were still looking in the plane's direction suffered permanent eye-damage, although most such unfortunates could not see well for more than a full day.

Within the group of squatting folk at Hakeem's turning-point, all attention, and all eyes, had returned to the speakers on the central platform. The uproar of conversation lasted a minute, perhaps two. Then, as Ramji raised his bullhorn to get attention and resume his talk, Daisycutter's N-17 detonated. Ramji and Hamid reacted identically, turning to Lakshmi and shoving her to the platform, then draping themselves over her as shields. But although appreciated by all, especially Lakshmi, the heroics proved unnecessary. Thirty seconds and the fireball was gone, cooling rapidly to insignificance: only a trivial remnant of the shockwave ever reached them.

Overhead, several nations' satellites spotted and identified the blast, reported it within seconds: even before the fireball had fully cooled, most of the world's governments and militaries knew of the event. Shortly they, and the public, would know a good deal more.

Down on the deck again, hiding from Indian radar, Hakeem throttled back to maximum speed without afterburner. He didn't know what the response time of the Indian Air Force might be under these conditions, but his personal job now was to either get out of his plane safely (probably to be beaten to death by a mob if he survived ejection), or (and this was by far the better idea) he could set Tango One down at some Indian Air Force facility before India could get planes up to intercept him. Once on the ground, he would demand political asylum – he had little doubt about it being instantly granted.

He knew that the closest useable Indian air-base was Barrackpore in West Bengal, a rather short distance south and west of ground zero. Planes with characteristics similar to the F-16's were based there, so their runways could certainly handle Tango One. He knew the base's call sign – VEBR – and chose the most likely control tower frequency. And got lucky – he made contact immediately.

Hakeem spoke tersely to the tower operator: "VEBR, VEBR – Emergency! Mayday mayday mayday. This is Pakistan Air Force F-16 Tango One, piloted by Col Hakeem. I am the aircraft responsible for the atomic explosion over the Delta. I disobeyed orders to bomb the Delta and I jettisoned my weapon as safely as I could. I am inbound to Barrackpore AFS at Mach zero point 98, altitude about 100 metres. I intend to land immediately on the longest runway available and surrender myself and this aircraft. I have no further weaponry on board. I am now dropping my remaining external fuel-tank over water, and will land clean. What are your instructions, over?"

The tower mike managed to stay keyed: Hakeem had a balcony seat at the frenzied, confused discussion which went on for perhaps thirty seconds before an authoritative voice took over. "Say again your name and aircraft, this is Brigadier Solnoy, commanding officer of VEBR."

"Former Col Hakeem here, General. I resigned from the PAF approximately fifteen minutes ago. I now formally seek political asylum in India. I disobeyed two levels of direct orders to bomb the half-million people in the Delta and instead disposed of my bomb out over the far eastern edge of the Delta. My aircraft is an F-16, now unarmed. I will land at Barrackpore from north to south unless directed otherwise. I propose a full-stop landing, after which I will formally surrender myself and my aircraft. Are there any runways which I must NOT use? I will have no time to inspect the facilities. How do you plan to handle my aircraft?"

After a few seconds, the Brigadier came back on line. "I will be honest with you, Colonel – this is a unique situation, without precedent or prepared protocols. I will have to handle this on my own authority – I have no idea what the powers that be in the Air Force or the national government will want to do, how they will choose to deal with this event. But meanwhile we will wing it. On my own authority I accept and grant your request for asylum. You may land here without opposition. All other airfield operations are now suspended, our airspace is closed. You may land on any runway, north to south is excellent, 160 left is our longest. Land to a full stop on the runway, then up canopy and wait, engine idling please. We will send a guide vehicle for you to follow: we have no ground equipment to handle your F-16, so we ask that you maneuver at our ground crew's direction but under your own power. Our "Follow Me" truck will guide you into an empty hanger, where you will shut down completely please. Can you exit your aircraft without a special ladder? Do you carry a personal firearm on you?"

Hakeem chuckled, his tension beginning to dissolve. "You, Brigadier, must be a pilot. I can get out of the plane without a ladder. I do have a Walther PPK automatic pistol – that's James Bond's favorite, you might remember - which I will unload and leave on the seat when I exit the aircraft. I will also disarm the ejection seat as I exit, so that your ground crews need not worry about it. General, I must now pop up briefly to 2000 m to find your runway. Please do not shoot at me!"

An uneventful landing.

Parked in the hanger, Hakeem shut down his engine, went through the "shut down to cold" checklist. Saluted the group assembled to meet him, and noted with puzzlement, and then with relief, the lack of weaponry on the greeting committee. Took off his helmet, stood, got one leg over the edge, dropped the pistol and its ejected clip into the seat, after clearing the chamber. Swung himself out of the cockpit, dropped to the ground, and then collapsed against the nose-wheel strut as if made of badly overcooked macaroni, suddenly boneless, exhausted, gasping.

Two men helped him back to his feet. He managed a salute to the Brigadier, who eyed him for some seconds, then said quietly "Were you really instructed to bomb the people on the mudflats?"

Hakeem nodded. "Sir, my flight plan and all instructions will be found in the aircraft's computer. Full documentation. Your intelligence people will undoubtedly have recorded the entire mission… as will have other friendly nations." Then, almost in a whisper, "There have to be several lakhs of people out there, General. Probably half a million, perhaps more. All they have is bare feet and head-bundles. Surely NO human being could do what I was told to do? Not a real human, at any rate."

"You disobeyed the orders, destroyed your weapon harmlessly? Did so despite direct orders to the contrary? On your own responsibility?"

Hakeem nodded again, said quietly "When I got that directive, it made no sense, so I checked. I broke radio silence, and I got direct specific verbal confirmation of the validity of the order, from a PAF four-star nicknamed 'Daisycutter' – to me, Brigadier, it seems likely that he is behind the entire business. He is well known to be unbalanced on the topic of India/Pakistan dealings – and he is dangerously aggressive at all times."

"At any rate, the verification made no difference to me - I could not do it. I refused absolutely, such an order goes directly against the Geneva Convention at least, and also against other, much higher authority. Among all the other horrors of that order, I could not imagine my name becoming the darkest curse pronounceable for the next many thousands of years. I have no family to disgrace, but THAT I could not do."

The Brigadier then stood to attention and very formally saluted Hakeem, who returned it, clearly puzzled.

"Mister former-Colonel Hakeem, because of your actions today, forever and ever your name will be a blessing, not a curse. Of that you may be quite certain. How many individuals in all of human history can truthfully say that single-handedly, on their own initiative, they saved half a million lives? You had orders to kill, and you had the means to do it. Yet you did not. You personally prevented the destruction of all those human beings. Further, you have undoubtedly prevented a nuclear war. The people you saved will likely never know that they owe their lives to you. But some of us do know. We will tell the world, believe me. And we will forever salute you."

The General dropped his salute, took Hakeem by the hand and said "Thank you for your actions. Believe me, no matter who the madmen are who planned this, given the state of alert of the whole world now, there is no chance for a second attempt – and I personally doubt any group would be able to access more than one bomb. That is, of course, until the god-damned idiot North Koreans increase their production and begin selling bombs at retail to all comers - with no questions asked. Now we have to find out what has happened here! I suspect there are some serious lessons to be learnt by everyone. And I also strongly suspect that your PAF career is over, that you are not going to return to Pakistan in the foreseeable future. Who knows, perhaps we can find something to interest you in this part of the subcontinent? We shall certainly have to see, won't we?"

Then with a huge smile "Come with me, Brother, to my office where we can have at least one cold beer before the higher authorities arrive from off-base and make all our lives even more miserable."

12

NUCLEAR DENOUEMENT: REWARDS

{T_0 + about 35 days}

For almost a month relations between Pakistan and India danced on a knife-edge, with a great deal of rattling of nuclear sabers, plus intense red-faced screaming and shouting resembling nothing so much as fourth-grade boys in a schoolyard, wrangling loudly over a dead frog or some other important item. On the third day, the Indian government decided that its public needed some sort of Roman-circus-style entertainment, and they ordered Hakeem's lovely airplane publicly burnt rather than returning it as the Pakistanis were demanding whilst claiming loudly "THEFT". The open-to-the-public burning was witnessed in person by a couple of lakhs of citizens: so resembling a funeral pyre, it did seem to calm public uproar somewhat. Bangladesh's role was glossed over and ignored, although it was bruited about that perhaps she had asked her former partner for help in dealing with the MSL-migration problem, and that this incident was somehow related to the request.

No nuclear war ensued. Why not? As we have just seen, small things, sometimes seemingly ridiculous things, can trigger or prevent major catastrophes. What eventually actually prevented nuclear war and got the parties to stand down, was the public revelation (anonymous but believed to have come jointly from the American and Russian intelligence services) that the two countries' nuclear-bomb-development programs shared an unwanted and very much unpublicized nuclear world-record. Neither had a very extensive nuclear arsenal – bombs numbered in the single digits on both sides. And each development program had produced something never accomplished elsewhere in the world – two test atomic bombs (per country) which refused to explode when asked. No other example is known of an atom bomb failing to detonate when told to do so. Ever. Anywhere. For any builder. The countries eventually settled down instead of going to all-out war simply because neither side could guarantee the functionality of its own nukes. Nonetheless, things came so close to Armageddon that the event precipitated new (and still-active) talks about nuclear disarmament.

Nobody has yet been able to find Daisycutter – live or dead, he is definitely and thoroughly unavailable.

● ● ●

The amused Gods roared as they called for more mead. Many among them were greatly disappointed – an exchange of such primal-force weaponry would have been an entertainment most worthy of every god's attention!

● ● ●

{T_0 + 35-50 days}

Within a say, Hakeem's exploit was world-wide front page news. The stories included his name and a bad security-photo from long ago, speculated on his career and motivations. Most stories included a big selection of theorizing about "Who What Where When Why and How?" by people who really didn't know what they were talking about. He was spontaneously hailed around the world as a hero of the highest possible grade. Hakeem could not agree – to him, doing what he did was his duty: there should be nothing extraordinary about a military man hewing to his duty. Nor about any human refusing such an order.

At first, he knew little about that public praise, for he was under tight security – his thorough, detailed debriefing lasted thirty days. During it he had no access to any form of media save a local newspaper with a few items carefully excised. He behaved properly, honorably refused the role of traitor – he resolutely refused to discuss anything that might be classified, but spoke freely otherwise. To say his caretakers handled him gingerly and with respect is a considerable understatement – after all, Hakeem was now India's favorite modern hero – a Pakistani in that role, for heaven's sake!

Towards the end of the month, the question of his future came up – clearly he had no future in Pakistan, where he would most likely be murdered on sight. Nothing obvious came immediately to mind – the Indian Parliament voted him a thoroughly adequate lifetime pension as a thank-you, but he had no plans or goals as yet. His future was, indeed, a puzzlement.

After the month's isolation, he had been told he would be freed the next day, to be taken wherever he wished for the actual release from custody, complete with new identity. His keepers were trying to make that release into a low-key event – despite a huge public demand to see him in the flesh. He agreed that secrecy was the best idea, if possible.

Four PM, a knock on his room door whilst he was watching on his new TV the now-old reports about his flight, the bomb, and the amazing lack of casualties. He'd already watched several detailed video segments on the event, plus documentaries on the MSL People and the MP, all of which he found fascinating. An orderly stuck his head into the room, said "Sir, you have three guests, shall I take you to them now? They are in the visitors' center."

Hakeem hadn't received any 'guests' other than military and intelligence people – as he'd told Daisycutter, he had no close living relatives. For the past two decades he had lived a more and more ascetic existence – but without completely withdrawing. Now he was seriously contemplating something along those lines. At any rate, he had no reason to expect any guests, and also no reason to be discourteous to anyone interested in (and capable of) finding him... not to mention their getting permission to visit. Interesting indeed.

He stood, shrugged, asked the orderly "Civilians this time, perhaps? Not the press, surely. And after a month. This is most curious."

The orderly shook his head at Hakeem, said "No Sir, not the press and not the military – not even intelligence. A lady and two men, all three clearly civilians."

Hakeem stepped into the visitor's lobby, stared at the three people standing there. A woman in a scarlet sari, quite elegant in a softly-stated way. A middle-aged man in white. A very young man, obviously still in the process of leaving physical boyhood but very much an adult by demeanor. The younger man wore a flaming scarlet Nehru cap – the older man a similar cap of khaki but with a red vertical stripe.

It took perhaps five seconds for Hakeem to realize who these three were, that he had seen them before. From a distance, but clearly, out there over his left wingtip. He had no idea what to do or say.

The woman finally broke the silence: "Sri Hakeem, I believe the three of US have ALMOST met YOU quite recently, under rather different circumstances. I am Lakshmi, this is my husband Hamid. This young man here is Ramji, founder and leader of the Migration Project, an important grass-roots movement amongst the MSL People. He is my husband's and my own student in high-school. We help him with the MP movement however we can."

Hakeem was deeply flustered and could at first say nothing, therefore busied himself shaking hands all around, finally managed to say "Um. Well, Sri Ramji, I believe you had a bullhorn in your hands last time we met. And you, Madam, were, I suspect, wearing that very sari at the time – most distinctive it is, even from an altitude of one thousand metres."

The four sat down at a table in a quiet corner.

Ramji said "I think it best to come straight to the point. Or points. First, the three of us are here at the request of the MP, as the personal representatives of the half-million souls you were sent to kill. And whom you did not. To say you have our thanks is the most amazing understatement."

Hakeem flushed, could only nod, then cleared his throat and managed "Every real human on this planet would have done the same. I did nothing one should not expect. But I am very glad that I did what I did."

Hamid: "We have been talking with your 'hosts' for the last couple of days. They tell us that you seem to have no specific plans for your own future – that you have not even yet picked a place for your release, and that a major consideration is maintaining some semblance of privacy for yourself amidst the obvious notoriety your person carries with it."

Hakeem shrugged, said "I have no relatives, no family to go to, not in Pakistan and certainly not here. My handlers' statement is true, I have not chosen a place. What do I know about such business?"

Lakshmi: "Then perhaps you would come with us? As our personal guest –after all, like it or not, YOU are our personal savior! You would also be the the guest of the whole MP --- have you heard about the organization?"

Hakeem flashed his first relaxed grin – "Yes, I have, on the TV they finally gave me yesterday. An amazing movement, fueled and generated entirely by the mud-flat people themselves – with you three leading – you are called 'Triad', I believe?!"

Lakshmi nodded, smiled, spoke again: "Please, consider being our guest, the whole movement's guest, at least for a while. It would be no imposition. You have no family- but you surely need a friendly, private place to stay while you think about the future. Hamid and I have a house much larger than we need, in a quiet mostly rural setting near the Ganga-ji. We have three fine rooms completely unused, any one of which could be yours. We live close to the school in which we teach, on the high south bank of the river – Ramji lives on the far side of the water some kilometers from school, out amidst the MSL People for whom he is now the chosen leader. Staying with us is an easy decision, if you choose to make it so. Come with us, see what we do, take time to think and consider your path… there is certainly no hurry. But you do need a place to call 'home' at least for a short while."

Hakeem was taken aback, but intrigued. And he had nothing else in the frying pan. As they said, it could be a very easy decision. He made it.

"Accepted, with thanks. I will try not to be a burden. I have received a nice Indian pension, so I will have plenty of money to pay for my needs…"

Said Hamid, firmly, "No talk of money, please. That is the ultimate in trivia for the moment, for reasons which will shortly become clear. We are grateful to be able to begin a tiny repayment on our debt to you. We will expect you to stay as long as you wish."

He stood, then all stood.

"We will meet you here tomorrow at four –you can discuss this plan with your local authorities if you wish, but they have already agreed to it. Wear sunglasses if you have them, a mild but quite effective disguise. We will stop at a clothing store and any other place you wish enroute home. And I expect our cook, Noor Mohammed, will produce a dinner significantly different from your local fare here."

Hakeem met the trio at 1600 next day, with sunglasses and ill-fitting new government-issue civilian clothes, money and new ID cards in his pockets, a folder full of papers containing the information needed to set up his new life when he decided what that meant. All contained in a cheap suitcase and a small backpack. Nothing more.

On the way home, with Lakshmi driving, Hamid broached another subject: "Sri Hakeem, do you yet know about the website with your name on it?"

Hakeem looked totally confused, said as much.

Ramji laughed, told him "The day after the bomb, a web site appeared to which people can make donations of cash – the recipient is YOU, all by yourself. It is a bit like crowd-sourcing for investments, except the money is entirely 100% yours. It is 100% tax-free, too, inasmuch as it is all gifts. The donors are from all over the world - just saying 'thank you' in the only way most of them ever will be able to."

Hakeem was speechless. Hamid turned in his seat to face Hakeem, patted his arm and told him "The sum is up to well over 160 million US dollars already, an infinity of Indian Rupees, all cash, all being held in trust in your name. There is incoming about two to three millions of dollars daily, and it is not yet slowing down much. You are suddenly a rather rich man!"

When he was finally able to speak, Hakeem muttered "What in the name of all the Hindu gods shall I ever do with such a fountain of money? I have no need or desire for it." Then, "Perhaps it is Allah's way of testing me further?" He snorted derisively: "Allah is not someone or something in which I believe, but is a useful metaphor. I shall have to find something constructive to do with that much money. Perhaps I will not spend it all in one place?" He laughed, squeezed Hamid's shoulder and said "Here I am, on the road to a new home after a big adventure, and instead of peace and quiet I get from my hosts such news as this – I am now custodian of a fortune! One which I did not earn, too. Rather like a rajah of olden days being born into his wealth."

Dinner was relaxing, quiet, and for the four of them only, very much to Hakeem's relief. Midway through the meal, Ramji said to Hakeem, "I do not mean to push, but believe me, there is very little to do in this town – in five minutes' walk you will cover the entire business district, there is no cinema, the library is in the high-school and not very much worth browsing except for the free copies of the Delhi and Madras newspapers always two days late. If you are interested in what we are doing at MP, perhaps you might accompany us three to my own home across the river, on the mud flats? There really, truly is nothing to do here except meditate."

Hakeem nodded assent: "A good idea, such a trip... believe me, all of you, the view I had from the air of MSL People on the march was impressive in the extreme. It looked very organized, from a distance anyhow. Your organization has undertaken a truly difficult task – even experienced, well-funded government agencies usually cannot handle such a need. Very impressive – I would like to see it more closely. When might we go?"

Everyone looked at Ramji, who thought for a moment and then said "You should have at least a day to just sit in the shade, or go walking about. Perhaps day after tomorrow, Saturday, when there is no school, we can all go over early in the morning and spend the day?" Murmurs of assent all around. "I will go home tonight in my own little skiff, return very early Saturday with the Three Brothers' large boat. My sisters will feed us lunch and dinner, no problem. Acceptable?"

It was.

After dinner that first day, Hamid stopped Hakeem for a moment, said with a huge grin, "Do you know that you already have a popular nickname?"

Hakeem looked embarrassed, then said "No, no idea… may I ask what it is, or should I perhaps not be told? It could quite reasonably be something most unflattering!"

"Hardly so" said Ramji – "You are known throughout the country as "Second-Sun Hakeem" – after all, you put a second sun in our sky, and without hurting anyone. The perfect stuff of future legends, Hakeem! And of course you have no choice whatever."

Ramji spent a very, VERY busy Thursday evening and all day Friday – getting word out up and down the Delta that Sri Hakeem, Second-Sun, would be arriving. An open, blanket invitation had gone out to all of MP to attend. A mere half million people got the word, people almost all of whose lives Hakeem had personally saved. His act had instantaneously established him firmly amongst the Hindu pantheon, where there is always room for one more. Most MSL People had never beheld a genuine god-in-the-flesh – this was their personal chance to do so, and not to be missed!

Early afternoon Friday, people began gathering around the podium, which had been moved farther away from the village, well out onto the flats themselves, to make room for the crowd. Arriving either afoot or by fishing-boat, they streamed into the grounds, filled them solidly for hundreds of metres, a roughly circular blob of white, always in gentle motion, expanding outwards by accretion, as if it were separately alive. The influx continued all night, into the morning. Fully the crowd-equivalent of three very large soccer stadiums completely full of fans. Perhaps even more – but counting them was unimportant and also impossible.

Saturday at 0600, the Brothers and Ramji arrived at the school, picked up the three other passengers, turned and headed back. The first two thirds of the trip were spent with Triad explaining MP to Hakeem, who was clearly quite interested and had done some homework, for he asked a flood of pointed, intelligent questions.

From two kilometres distance, it was obvious that something very, very big was going on ashore. The mud-banks and sand-flats for hundreds of metres around the stage were packed with people to the point of no ground being visible. Hakeem watched the approaching bank with an initial curiosity that turned to consternation as he realized the blanket of white was white cloth covering human bodies. Live humans all of whom very recently could so easily have been dead at Hakeem's hands - but who had not died.

The boat ran its bow ashore, the plank appeared: Ramji's hand-wave magically parted the white sea so that the four could walk to the raised platform. Hakeem was almost unable to take in the crowd, the entire event. Ramji walked with him arm in arm: the four mounted the podium. The huge crowd stood almost perfectly silent, staring. Ramji took the microphone, turned the system on with a loud click, then raised high Hakeem's right hand, and said to the crowd, as an opener, "This is our friend Hakeem, Second-Sun. The man who refused to bomb us a month ago. Refused to kill us all at one blow, which he had orders to do. But he did not, would not, could not do so. Our good fortune, my MSL People – good fortune beyond belief."

He paused, a long, drawn-out and pregnant silence. Lakshmi and Hamid silently admired his innate timing and delivery.

"And yet this man claims to have no family in this world."

Another pause.

"I would insist that he is completely mistaken... because ALL OF US HERE, EVERY LIVING MSL PERSON, WE ARE HIS FAMILY!

Then, quietly, "Am I correct? TELL HIM!"

The response was heard clearly at the school, across some eight kilometres of mud and water; it lasted for nearly ten minutes, and only stopped when Ramji himself turned up the PA volume to MAX and insisted on a return to calm. Hakeem was briefly in tears, composed himself, asked for the microphone, and made his first public speech.

"Thank you. Nobody has ever had such a family – I am honored. I am also impressed by what I have seen and heard about the MP that I assume you are all involved with – I do see several lakhs of those famous caps out there!"

"I have learned some things in my life- among them, that if one gets the attention of the authorities in any way that is not too obviously criminal, then one must be doing something important, and also something that frightens the authorities. That must be true of your MP - otherwise why try to bomb you? You are doing something important and doing it RIGHT. People around the world are sending me money for having done my duty to humanity. I do not need and do not want the money for myself – but I have now learned a great deal about your project. I have even seen it in action from the air. That program has a serious need for funds. I have decided that THAT is where the money is going to go." He looked at Triad, thoroughly enjoyed the shock on their faces. They knew the sums involved, both supply and need, and a balance could be struck, a fine balance, easily attained.

The cheer re-started, wave upon wave of sound – and when Ramji took off his scarlet Nehru cap and placed it on Hakeem's head, the roar increased amazingly.

Section Three – Freshwater vs Gulf Stream

13

JAYHAWK IMPRISONED

{T_0 + about 3 days}

Jayhawk's situation, not to call it plight, had grabbed the world's attention for a few hours, until public interest faded in pictures of a ship obviously going nowhere and not in immediate peril. Much more attention was, quite appropriately, being paid to the tens of thousands of people who had disappeared from the west coast of Greenland. Not a single body had yet been found.

Having no good idea how long their enforced stay might last, Jayhawk had immediately established a stringent, conservation-driven routine, and set the vessel's internal environmental conditions for an indefinite but long stay. Best estimates were they had something like 180 days worth of supplies – food would run out long before diesel fuel, given that the main engines were shut down. Some of the crew had already begun fishing for salmon.

The scientific party's radio link installed atop the nearby mountain handled all the communications needs of the ship. It made instant satellite-photo coverage continually available, which meant that there was no need to fly a Meadowlark to see what was going on outside the fjord – namely the steady southwards flow of very-large-scale ice-debris from the failed portions of the GIC. Jayhawk was imprisoned due to that flow, plus the obstinate refusal of the ice blocking the fjord's entrance to budge. Even had the blockage slipped away, the ship was still unable to sortie because in the massive flow of ice past the entrance there were no spaces big enough for a walrus -let alone Jayhawk- between the enormous floes. Consensus between HQ staff and the Captain was that if and when the entrance were to become clear, it would be unsafe to venture out without a great deal of open water for maneuvering, given that any random pair of even the smaller floes could squash the ship between them like a mouse on a freeway. After considerable discussion with CGHQ, Jayhawk's operational orders were amended to allow sortie at Captain's option when conditions reached 50% ice-coverage all the way from the fjord's entrance to the unaffected ocean perhaps 50 miles from shore.

The hoped-for "self-clearing" of the ice was well underway, the southerly Greenland Current providing free transport and exit for the flowing ice and freshwater. The ice's speed had slowed as the freshwater flow had slackened, but freshwater was still pouring out at The Boil at an astonishing rate: the uppermost several tens of metres of the East Greenland Current was for the moment 100% freshwater, at perhaps one quarter of a degree above freezing. That temperature is important - freshwater has its maximum density at 3.96°C... when colder than that, it expands (i.e., becomes less dense) and floats atop any warmer water. That propensity to float can have enormous consequences.

Over on the west side of Greenland, across the island from Jayhawk, most of the loosed GIC freshwater and ice had entered the sea in a single monumental torrent, and was now afloat and headed south, parallel to Greenland's west coast.

At the southern tip of Greenland, the eastern and western freshwater/ice flows merged, wishbone-like. That merged flow first turned southeasterly. There, it floated atop ordinary North Atlantic seawater and began to ride the North Atlantic's general, clockwise surface circulation, tending northerly at first, then turning eastward towards Europe. Satellite photos showed the flow was already busily discharging both ice and the two commingled huge puddles of freshwater into the northern reaches of the Gulf Stream.

14

JAYHAWK: SCIENCE PLAN

{T_0 +- about 2 days}

Rebecca was deeply frustrated – Jayhawk was the only ship in the world, not even to mention being the only research-oriented ice-breaker, in the right location to study both the physics and the biology of this wholly new phenomenon. She and her little team of students were literally the only oceanographers in the world with such an up-front seat... there was no other available research ship within three weeks' striking distance of any edge of the freshwater puddle. And certainly no other research-equipped vessel with a complement of oceanographers aboard. Yet here they sat, trapped indefinitely in this life-saving little fjord. So far, the ice filling the entrance had shown no interest whatever in moving: the suspicion was that the blocking ice was hung up on the "guardian rock" just outside the entryway – the rock about which Jayhawk had pulled her blind, life-saving full-speed turn into "Jayhawk's Fjord". Repeatedly documented by satellite photos, that door-ice had so far stayed put –not a wiggle- whilst the vast stream of floe-ice scooted on past at two knots, headed southwards into warmer climes.

Rebecca had a rough mental outline of a plan for the scientific work. Essentially, once freed, Jayhawk would dedicate as much time to Rebecca's physical and biological scientific programs as possible, would stay out at sea doing this GIC work to the limit of her resources.

Hoping for a bit of good luck (meaning, being able to take Jayhawk back out to sea soon) detailed contingency plans were developed to maximize scientific results if and when it became possible to sortie Jayhawk. Their situation, and numerous 'what ifs', had been discussed at length with Coast Guard HQ, with the funding agencies, and with the scientific community. The entire world-wide swarm of major physical-oceanographic egos was trapped ashore and had no choice but to leave the ultimate scientific program in the hands of a scientist who although nearly a "certified oceanographer" was nonetheless "not-quite" ideally qualified – in some folks' eyes, anyhow. Egos aside, all agreed that the opportunity to closely study this event was not to be missed. A plan was quickly developed which addressed the poor inventory of scientific equipment aboard Jayhawk, and considered what **could** be measured (vs. what measurements it might be nice to have). Plus details of where to do the observations, how closely spaced in space and time, how to vary sampling based on early results – all prepared as friendly advice to Rebecca, not as 'orders'. She welcomed the input, and particularly on how to extract maximum information from the ocean, using the miscellaneous and mostly antique equipment available.

Advice, yes – directions, no. A call had come through from her major professor, inquiring as to plans and progress – and during the conversation he made it quite clear that although the press had discovered their "student/mentor" relationship and were pounding on his door, this was HER show. He would vouch for her – enthusiastically- to the press and any others who inquired as to her training and abilities, but in his view she was now doing what she was being trained to do – think and act independently. His statements to the press would amount to saying he felt she was superbly qualified and that he could not see any reason she might need his, or anyone else's help. He left her with an enormous warm feeling deep inside – accompanied by an edge of near-terror. The affirmation was soul-filling, but the actual 'being in charge' was daunting.

The three highest-priority scientific questions were agreed upon – (1) the rate of movement and areal extent of the freshwater puddle; (2) the thickness of The Puddle (a vital measurement not possible from space); and (3) how fast the bottom of the freshwater layer was being mixed with seawater, down about 50m deep, at the fresh/salt interface.

The first could be readily and accurately mapped using satellites – the freshwater puddle showed up spectacularly well in infra-red imagery, and also in color pictures taken at green wavelengths, due to a complete lack of chlorophyll in the fresh water. Jayhawk would provide ground-truth measurements to ensure proper interpretation of the photos. The second would be easy to get by yo-yoing the football in the upper 100m or so.

But number three, the rate of mixing (i.e., turbulence at the bottom of the freshwater layer) was a much more difficult thing to get at – shipboard measurements are difficult, and no good way yet exists to infer small-scale subsurface turbulence from satellite observations. And especially not turbulence occurring down tens of m deep.

With the miscellaneous instruments and gear now aboard Jayhawk, turbulence (a.k.a. mixing) could be neither directly measured nor easily inferred from other data. Rebecca's original biology-oriented program (pre-GIC) had not included studies of mixing rates - which are among the more difficult things measured by oceanographers. She was anything but an expert on turbulence - not to mention that she had none of the specialized turbulence instruments. Nonetheless, extensive discussion with mixing experts ashore had produced a plan that should get reasonably good, and quite valuable, "approximate" mixing data using the instruments available. Most important would be her football; properly yo-yoed up and down through the interface where freshwater and salt water met, its sensors for temperature and salinity could provide the data needed for coarse estimates of mixing rates – coarse, yes, but infinitely better than nothing. There were plenty of competent people aboard to help with the program 24/7. It would take a lot of "instrument wet" time, but so long as no instrument-catastrophe occurred, it was a simple task.

Of course, instrument capabilities aside, the desired science could happen only IF the door opened and IF the ice-pack loosened to a safe degree.

{T_0 + about 12 hours and onwards}

Meanwhile, aboard Jayhawk, the news channel's content was dominated by some god-awful idiocy over in India. The story was garbled and continually evolving, with confusing loops and backtracking. The core of it seemed to be consistent – an atomic bomb had gone off in the air over the outer fringe of the Ganges Delta, in international airspace, amazingly enough without damaging anything or killing anyone. The F-16 and its pilot, who had delivered the bomb, immediately fled the scene and landed in India. He was Pakistani, apparently acting under properly-issued orders to bomb half a million penniless Indian refugees who were fleeing from the Ganges' flooding – which order he had chosen to disobey. He had at once –without question- been given asylum in India, become an instant international hero, and then disappeared from view.

Much less clear were answers to some obvious questions… WHOSE bomb, and WHY? The Indian and Pakistani governments seemed to be the only logical suspects, but were keeping remarkably mum. Rampant news-pundit theorizing held that a bomb had been stolen (yes, a reasonable possibility, but from whom?) and had then landed in (obviously!) some very wrong hands. Hands with access to Pakistani Air Force F-16s, in fact. While any number of well-organized terrorist organizations could have orchestrated the mere theft of a bomb, no known terrorists owned and operated F-16s. A puzzlement indeed.

As a sidebar that seemed entirely unconnected to the bomb, the news channel aired a three minute interview with an Indian oceanographer. He discussed the massive human migration that had been underway on the Delta when the bomb went off – and connected the migration to the abrupt (but small) rise in sea-level due to the sudden, massive input from the GIC. That rise, he explained, the Delta's resident population had widely misunderstood as a harbinger of a great deal more to come immediately. That misreading had apparently touched off the panic-driven migration so dramatically shown in aerial news-footage from the scene. Jayhawk's crew found it endlessly intriguing, this connection between their own predicament, and that of half a million mud-flat residents, half a world away, needlessly (at least, for the nonce) on the run for higher ground.

It would take months of careful data-gathering and analysis by several agencies and news organizations to establish the linkage between the bomb and the Ice Cap meltdown.

The Gods, however, already understood – and were hugely amused.

15

JAYHAWK: INFORMING THE WORLD

{T_0 + about 3 days and onwards}

Day One of Jayhawk's imprisonment. Normal satellite communications had been established within hours of Jayhawk's escaping into the fjord: Rebecca had immediately reported in with family and colleagues. In the afternoon and early evening she had been preoccupied with making detailed plans for studying the freshwater Puddle (with "Puddle" now a proper name, hence capitalized even by the press) as it moved out into the North Atlantic. That plan of course assumed that Jayhawk would get loose before some other ice-strengthened research ship could get to the region and steal the 'Hawk's thunder.

Tired, but with the day's uproar behind her, Rebecca stood alone in front of a brand new thermal satellite photo overlaid on a chart of the North Atlantic. Both the Gulf Stream and The Puddle showed clearly – the Stream as a series of more or less parallel bright-white stripes of very warm water. A perfectly normal image: unlike terrestrial rivers, the Stream has no defined channel, and often divides into multiple sub-streams. It wanders about rather chaotically, hence its location for tomorrow cannot be predicted with any certainty, regardless of how long it may have sat in today's position. The Puddle stood out brilliantly – it was at just about zero C, much colder than the Stream's surface. In part of the western North Atlantic, the warm-water stripes had already disappeared under the advancing flood of freshwater and ice – and that was with most of those materials still queued up for entry into the North Atlantic.

This was her first good, focused "grand overview" of events. She was considering how the masses of freshwater and ice from Greenland's east and west coasts were coalescing at the south end of the island, then heading mostly eastward tending northerly. The freshwater and ice were being simultaneously floated from below and pushed from behind, riding atop the Gulf Stream and the North Atlantic's general clockwise surface circulation.

Rebecca flipped idly through a stack of hard-copies of several successive hourly images. The sheets were clipped together at the top, and her ruffling the stack with her thumb made almost a movie of the progress of The Puddle. It was in the third cycle through the 'movie' that it hit her, rather like a punch to the solar plexus. There was a big, BIG problem looming. She knew that the GIC had dumped only two percent of its water – enough to raise global average sea-level about fourteen cm, a figure everyone aboard had at some point calculated or been told. For some places on the planet, that would be a big change, producing all sorts of very strong local effects. The many mid-Pacific atoll nations with freeboards of ten cm were in deep trouble, starting some years ago, and this sudden increment wouldn't help them at all. Other places, +14 cm would be a big 'So what?'.

"True enough…" she mused: "…the GIC water, if spread evenly over Earth's ocean, would make a layer 14 cm thick."

She kept going. Thinking "But the materials are NOT going to be spread thinly over the globe!" Not at all. She zoomed in on the important realities. From the "flip movie" one could see immediately that the GIC's freed water wasn't going to spread across the whole of Earth's ocean very fast – rather, it was going to be concentrated into the far-northern Atlantic, where ALL the GIC water and ice certainly seemed to be headed. There, it would cover quite deeply a relatively tiny percentage of the world ocean… tiny, but critically important.

Corralling the freshwater into a relatively small area would change things dramatically, by concentrating local effects: she realized that the number everyone was focused on, namely "+14 cm on average", was meaningless.

She used an old wooden ruler on the chart to get approximate areas, grabbed her calculator, and did the new calculation, for the area of the North Atlantic that would receive the GIC 'debris'. She did it three times. With even vaguely reasonable numbers as inputs, the outcome was disaster. At least 50 METRES of freshwater... perhaps even more. Freshwater and ice would cover the entire breeding ground of the Gulf Stream at least tens of metres deep.

Rebecca was aghast. She knew the physics of world ocean circulation, and an obvious fact leapt out, making her shiver.

Disaster.

The floating GIC fresh-water was going to kill the Gulf Stream.

Stone. Cold. Dead.

The disaster would produce a sudden, radical and probably permanent change (at least on human scales) in the climate of Great Britain, much of Scandinavia, and of western Europe. As soon as she identified it, the disaster seemed inevitable. And most certainly not subject to human intervention. (The Gods, of course, had been forbidden to interfere in the game.) Any oceanographer could quickly figure this out. After all, the concept of 'GS shutdown' wasn't her invention - the possibility and the likely mechanisms had been discussed in her introductory physical oceanography course years ago. But SHE knew what others differently trained either didn't know at all, or chose to ignore, namely the likely ecological and societal consequences of the freshwater now enroute to the central North Atlantic.

Before she could clear her mind from the shock of the idea of the imminent death of the Stream, the intercom crackled, and one of the bridge watch-standers came on. "Doctor Rebecca, you there in the comm hut?"

She grimaced wryly and acknowledged the hail – it was swimming upstream to get the crew NOT to call her 'Doctor'.

"Well, Ma'am, we up here on the bridge –meaning the Captain- just got a sort of mystery phone-call, someone trying to get ahold of you. Shall I patch her through?"

Rebecca wouldn't find out until much later, but in their scramble to find someone –anyone- both authoritative and "on-the-scene", reporters covering oceanic aspects of the early hours of the GIC disaster were at first constrained to doing their own interpretations of generic satellite photographs… a difficult task for the unwashed. One free-lance member of the press corps, an aggressive young science reporter named Anita, quickly discovered that there was supposedly an articulate, knowledgeable, and personable (not to mention FEMALE!) oceanographer aboard the trapped icebreaker – as close to "on-scene" as one could imagine. And undoubtedly the only oceanographer in-situ. A potential scoop, at least a limited exclusive, if she could work things.

Anita was hoping to generate a good, lasting contact with whom to work up first-person-ish stories. She came from a military family and understood chains-of-command, and protocol, hence she had first called the ship's Captain, who had been unreservedly enthusiastic, and had ordered the connection transferred directly to Rebecca.

"HER? As in FEMALE?" Rebecca thought, then said "Sure. Any idea what it's about?"

"She said something about being a science reporter. What's happening to Jayhawk is probably a hot topic with people like her. I'd bet money she wants an interview! Here you go, Ma'am. Good luck!"

Rebecca gritted her teeth: the 'Doctor' business had a parallel irritant - she positively hated the way she'd apparently suddenly become an ancient, just by boarding the ship. "Ma'am" indeed! Much more irritating, however, was the idea of a science reporter… Rebecca and several of her graduate student colleagues had gone through unfortunate experiences with that profession. Reporters seemed positively incapable of getting scientific things RIGHT – a widely-recognized serious burr under the saddle for scientists generally. Nonetheless, she took the call – aware that she probably would not have done so for a male reporter.

"Rebecca Wilson here, on Jayhawk" she said. "Who's calling?"

The caller spoke rapidly, as if afraid of being cut off. "My name is Anita Jantzen, Dr Wilson. I'm a free-lance science reporter, with good contacts at several of the major TV and cable networks. You folks' predicament made headlines and piqued my interest. I got a copy of your crew-list and your overall cruise plan by calling Polar Programs at NSF, which sent me to the Coast Guard and here I am. Getting your cell and Skype connections was simple, but I thought it might be diplomatic to call your ship's Captain first – he just transferred me to you. If you are up for it, I'd like to interview you about what's going on with the collapse of the ice cap."

Rebecca quickly re-examined her distaste for reporters, realized that this particular individual must know more than most – going to Polar Programs had been a crafty idea. Therefore Rebecca didn't hang up – instead, she said "Ms Jantzen, you should know that I have a very strong aversion to giving interviews. In my experience the reporters are almost always scientific illiterates, and essentially cannot get things right. For instance, you just now referred to the 'collapse of the GIC' – which is not at all what happened. I don't mean to be preachy, but I was taught that language is best used with both precision and accuracy... mistakes will happen even without the author actively generating them! Sorry to sound so negative, but I've had some bad and embarrassing experiences. And by the way I am not yet "Doctor" – shortly, I expect, but that title is presently inaccurate. 'Rebecca' will do just fine."

A long pause. This time, it was Rebecca who thought perhaps Anita had disconnected, but no: Anita came back online. "Ahem. I understand, believe me. I'm not one of those illiterates, Rebecca – I have an MS in organismal biology and a BS in chemistry. Plus some years of this sort of work. I do understand your position on reporters, although often the crappola is not their doing – frankly, a lot of my time is spent fighting with editors over 'truth, precision and accuracy' concerns about stuff that goes out with my by-line. I'll tell you what - if you'll agree to an extended interview, via Skype, I will guarantee you a final chop on not just what I write, but also on the final form and content of what goes out to the public after editing. I'll depend on you to keep me accurate, hewing to your personal standards. Would that help?"

It was an unusual and quite intelligent offer. Rebecca was significantly mollified. "You say you've gotten ahold of the cruise plan: that means you know I'm not Chief Scientist for this cruise… that would be Doctor Adkins, the glaciologist who is studying the ice cap. I study the ocean, not islands covered with ice. My own bio program is a sort of independent tag-along to Doc's big ice-cap study. You should really be talking to him about the ice-cap. I'm certainly not qualified in any way to discuss the cap or its partial collapse. Me, I'm an ecosystems type, but I do know a bit about the physics of the ocean."

She stopped, thought, then continued: "I am, however, the nearest thing to an oceanographer aboard this ship, and I just a few minutes ago realized that there are some very, VERY serious biological and social ramifications to having that much freshwater input to the North Atlantic, all in one bolus. Perhaps we could go into those…"

Anita interrupted: "Don't worry – I'm clear about your field of expertise, but I understand that Doctor Adkins's entire ten-year project washed out to sea or was otherwise destroyed by this event. I was just told that he also lost all eight of his graduate students who were on the ice. I'm terribly sorry about that – god only knows how many Greenlanders were swept away, tens of thousands certainly."

A goodly pause, then "Rebecca, I understand pecking orders in both the military and scientific research. Therefore I tried first to get through to Adkins, but he's understandably not available, and his underlings are being stone-wall protective. Hence I'm turning to you as leader of the other half of the expedition. Besides, some of Doctor Adkins' own flock recommended that I talk with you regardless – they think your science is neat, and that you're a wiz at explaining things in laymen's terms, which is what I, too, like to do.

No need for us to preach to the already-sanctified – we need to get to the millions of unwashed neophytes. While also not allowing or committing too many silly errors and thereby pissing off the scientific community."

While Rebecca appreciated the compliments and the philosophy, ultimately it was Anita's incidental but correct use of the term "expedition" that settled the question of an interview, in Anita's favor... clearly she had some real understanding and savvy.

"Okay, Ms Anita, let's do your interview. But it has to be about oceanic effects of the Cap's collapse... NOT about the actual collapse itself. You need to send me a written signed statement about my having chops on our materials, to which condition I will hold you quite tightly. With that done, we can work together. Ground rules – we are NOT going to go into the GIC collapse. Period. That's Doc's purview and there will be plenty of off-site commentary and analysis of that event. What I CAN legitimately discuss is the ocean's response to the freshwater input. There will be very strong physical and biological consequences. Plus, perhaps we can go into some of the huge social implications of the collapse. A general audience is going to need a good bit of educating, so some of our time will have to be with me in "teacher" mode... maybe with you as my foil. What exactly are we talking about? A two-minute or five-minute finished product?"

Anita's voice changed – she sensed a different and perhaps more interesting story. "Before I answer that, would you tell me what you mean by 'huge societal implications' just a few seconds ago?"

Rebecca replied by asking "Anita, do you know what the Gulf Stream is?"

Anita said yes, she did.

Rebecca said "Good. You're in a small, select minority. How about this, stream of consciousness, top-of-my-head: The GIC is 100% freshwater. Low density. Floats on salt-water. As we speak, I'm looking at today's satellite photos centered on Jayhawk's position. It looks like the entire load of water and ice released by the

GIC will wind up in the central North Atlantic. That's almost certainly going to be a disaster of global proportions. The freshwater will shut off the motive power for the Stream, and the Stream will stop. We can't predict for how long - perhaps permanently. England's relatively balmy climate is caused by air warmed by the Stream. England will suddenly, literally overnight, become subarctic – no more grain crops. None of London's buildings had been constructed for the new climate. We have no idea at all what might re-start the Stream, or when that might happen... if ever. Is that enough of a teaser? I'm just sorting this out for myself, Anita. Explaining it to you and your audience would be a good exercise for ME as well."

Anita's news-nose was twitching frantically. She took a deep breath and changed directions completely from her initial intentions. Let the other reporters have the physical collapse... Rebecca's take on things biological and social could well be much more important.

"Good GOD! Jeez, Rebecca – this is a very different story from what I expected to write. It's a hot one, really important. We have to do it absolutely ASAP – it'll be a scoop, an exclusive. Can we do the interview this evening, your time? Skype will do just fine. My guess is this should go in as a 30-minute "evening news special detailed report". We'll assemble a thirty out of whatever materials you and I put together. Usually the producers want five or ten hours of interview from which to prepare a 30-minute program. This is pretty concentrated stuff, so let's you and me expect about three to four hours of interview time. Okay? The production crew I'm going to use is always on 10-minute standby, so they should be able to get the thirty together very quickly after you and I finish... probably in a very small number of hours. Like, single digits."

Rebecca was beginning to relax – Anita seemed eminently sensible, and Rebecca simply loved teaching complex science to the lay-public. Besides, the world needed to know what she now knew. She agreed: "Okay. Should be fun."

Anita observed, astutely, "It's going to be a good deal more than 'fun'! Rebecca, neither of us is stupid. You're a senior PhD grad student and the only qualified person in a position to study or even report on a unique new phenomenon in your own field. This HAS to be a career-maker for you, if you play it right. Likewise for me... this is a global story. If we can do a good job that gets picked up and re-broadcast, then I will be golden, too. Both of us together! Maybe you can guarantee me an exclusive on your time for a week? Sounds nutty, but it would be a good deal for you – give you a shield against all the other reporters who are going to come piling down upon you... believe me, if I can find you, so can others. And they will. You can just tell them "Sorry, folks, but Anita has an exclusive.' I guarantee they'll go away immediately."

Rebecca was quite pleased – no trying to cram an hour of material into a standard 140-second block between commercials! "Thirty minutes of air-time!? Wow! Not many of those come along. I accept your idea of a week's exclusivity, just put it into the note about my editorial chops. Let's see... I can outline a talk-slash-lecture in half an hour, give me two, better three hours to prepare a few vis-aides... graphs, satellite photos. I'll have our techs set up a video camera here for showing maps and diagrams and so forth. Let's get back together in four hours, ready to roll. You call me direct on Skype. I'll play teacher/scientist. Your role is interviewer but also student. We'll do the interactive education-plus-news thing. We can stay online together for as long as it takes. Reasonable?"

It was. Ten minutes later, the signed document arrived on the bridge, was carried down to Rebecca at once – she was satisfied, but careful. She spent some minutes checking Anita's bona-fides (excellent), then consulted Captain Shelton just to be certain of his okay. He was enthusiastic, thought the interview a fine idea and saw no reason to get approval from HQ. "What the heck – we ARE the only technical folks on-scene, we're doing nothing classified, immoral or fattening, and anything we can tell the world might just help folks understand what's happening. Go for it! The Coast Guard can withstand almost any barrage of favorable publicity – just don't slather anything on too thick!"

Things came together quickly. Rebecca did a quick outline for the lecture, emailed it to Anita as an interviewer's cheat-sheet. Then she scurried about for maps, diagrams and data. Everything she needed was simple and readily available on-line. She set up a relatively uncluttered, well-lit space in her lab, got her electronics tech James to jury-rig a switching system and video camera so that she could easily send imagery of graphs, satellite photos, and suchlike. When Anita called back as scheduled, Rebecca was ready.

To start the interview, Anita introduced herself and Rebecca - with emphasis on Rebecca's official status as a senior graduate student specializing in things biological in the ocean, and definitely NOT an ice-cap expert. Anita briefly covered the partial collapse of the ice cap, explained that the collapse itself was NOT the topic of this program, which would deal with various consequences of the collapse. In turn, Rebecca then explained briefly the overall plan of Jayhawk's cruise, her own scientific program and background.

Then, on Anita's cue, Rebecca began the real interview: "A big chunk of Greenland's ice cap has collapsed into the ocean over a mere ten hours. About thirty thousand cubic kilometres of water and ice wound up in the ocean. That's a lot of water, even though it's only about two percent of Greenland's total. We don't yet know if the remaining 98% is going to stay put. If the rest gets dumped into the ocean, it'll change sea level by over seven metres – that is, over 23 feet."

"Today's input of water from Greenland is 100% FRESH water it contains no salt and also no life – the water is sterile. Because it's fresh-water, a lot of interesting things are going to happen to the ocean and to humanity's interests in the ocean. Unfortunately, none of those things are likely to be good. To understand what's coming, and how fast, we need some background on the ocean and how it behaves. Especially the North Atlantic Ocean, which is where the materials from the Greenland Cap are going to end up."

She proceeded with a quick overview of the Gulf Stream, explaining to Anita and the camera. A good chart of surface currents of the whole North Atlantic. The Stream itself in that context - size, location, behavior. After describing the general circulation of the Atlantic, and connecting it to the global ocean system, Rebecca told the audience "Folks, the take-home message is that the Gulf Stream is an important part of a very complex four-dimensional flow of water that goes on all the time in the North Atlantic. That flow closely connects the North Atlantic Ocean to all the others. The overall multi-ocean flow, with the Gulf Stream included, is often called something cute like "The Great Oceanic Conveyor Belt". Especially for human affairs, the GS itself is extraordinarily important."

Anita fed her an open-ended question: "The Gulf Stream certainly doesn't look like it's a huge part of that overall circulation, Rebecca… so just exactly WHY is it so important for us humans?"

"Anita, everyone should look at a globe at least once – now would be a good time. Earth really truly is NOT FLAT even though our maps of it are flat. Flat maps of a spherical object are always wildly inaccurate." She held up a simple twelve-inch globe. "Here," she said – "…we can use this little fellow to illustrate the Stream's role in both the ocean and the atmosphere."

"In fact, the Gulf Stream is a major player in keeping much of Earth's surface habitable for us fragile humans. Look carefully at any globe and you'll see that most of Earth's surface-area lies in the so-called 'mid-latitudes' – say, between 50°N and 50°S - which is also where over 95% of humanity lives. Look at my globe – my fingertips are at 50° north and 50° south. Look at the area covered, It's HUGE! That much surface soaks up a lot of heat! Here's the simple physics of sunshine and Earth. During daylight hours all of the mid-latitude surface gets a great deal more solar heat, from sunshine, than it can dump back into space overnight through simple nighttime cooling. This means that the entire region will accumulate heat – get steadily hotter – unless some mechanism carries heat FROM the mid-latitudes INTO regions where great quantities of heat can be radiated out into space. If, on average, incoming is greater than outgoing, then you get hotter – no choice whatever."

"So with all that excess input of heat at mid-latitudes, why isn't Earth steadily heating up, and FAST?" asked Anita.

"Because we have radiators. Luckily, Earth has two regions that can radiate a LOT of heat back into space – all the area in the Antarctic and Arctic, north or south of latitude 60, has outgoing heat greater than incoming. Those high-latitude, near-polar 'radiators' are the Earth's counterbalance to the excess input at mid-latitudes. But folks, think about this - if you bring the excess IN at a different place from where you DUMP it off the planet –and that's what happens here, for Earth- then you must move the heat from the actual 'input' location over to an 'outgo' location if you're going to get rid of it. So here's the rub - the Gulf Stream is a major mechanism for moving heat from the "excess input" regions to the northern "polar radiator" region… the Arctic Ocean. Plus the sub-arctic part of the North Atlantic. Here's a jargon alert – 'sub-arctic' just means 'damn cold much of the time, but not quite arctic.' If Earth didn't have the Gulf Stream, things would immediately go whacko with the planet's air and ocean temperature arrangements."

She kept on: "Now bear with me, because we're getting to the meat of things. The Stream is composed of a huge amount of water, moving at a very good clip. The Stream needs an ENORMOUS amount of energy to keep it moving. Something has to shove all that water!"

Anita stepped up: "So, Rebecca, just where does that energy come from? Presumably not from the sun, or from Earth's core, or something equally weird!?"

"That's a good question, Anita – and the answer is a little complicated and a whole lot of odd. So, now for some very simple chemistry mixed with some unexpected physics." Then, directly into the camera, "Once we get the Gulf Stream working here, we'll talk about what happens if and when the Stream DOESN'T work. It is known to have stopped and restarted many times – we can see it in the geological record under the Stream's path."

"Back to energy. The major energy source for moving the Gulf Stream is a strange wintertime phenomenon in the North Atlantic. North Atlantic surface water is merely ordinary oceanic salt-water. Salt dissolved in water yields a sort of anti-freeze, and unlike freshwater, salt-water has no specific freezing temperature. But of course Mother Nature **does** make ice out of salt-water, every year, at or near both poles. Here's what happens. It involves salt and density and gravity, so it's a mix of physics and chemistry – but it's not very complicated and the result is amazing. In winter, ice crystals form in the very surface of the water, next to the super-cold Arctic atmosphere. The crystals are made of pure water and are microscopic at first. The important thing is that the WATER freezes, but leaves the salt behind in the still-liquid water surrounding the new ice crystals. That "excess" of rejected salt mixes with the water immediately around the forming crystals: this makes the remaining salt-water even saltier. The extra-salty water is called 'brine'. This happens pretty much non-stop for several months every year, over most of the North Atlantic."

Rebecca took a deep breath and charged ahead: "Here is the physics. Whenever you have a denser fluid on top of a less dense fluid, the denser material will sink. Pour dense cream into coffee and the cream sinks, puddles on the bottom of your cup – or, put another way, the coffee floats on the cream. Until you mix the two, that is. This production of dense brine is not an imaginary or theoretical process - it is easy to take underwater photographs of it in action."

A longish pause, then "The kicker to remember is this - the sinking of the dense salt-water produced at the surface, in the North Atlantic, drives the entire "conveyor belt" - and the conveyor belt's included child, the Gulf Stream."

Anita sputtered momentarily: "You mean, Rebecca, that all the energy to make that huge current GO comes from such a simple thing?"

"Yep--- that's correct. The process goes on all day, all night, for months on end, over between five and ten million square miles of ocean. The cumulative effect is a steady downward movement of very cold, very salty, water. The vertical movement of cold dense water, driven by gravity, both produces and powers the Gulf Stream. In fact, that sinking, dense seawater BECOMES part of the Gulf Stream."

Anita: "Wow. Talk about unexpected! I am purely amazed that such an apparently small mechanism can cause so large an effect!"

Rebecca: "Now we're going to change scales – move from microscopic for brine production, to thousands of kilometers for the overall Gulf Stream."

She brought up a chart of the North Atlantic, with currents.

"As the Stream flows from east to west through the tropics enroute to the southern end of North America, it picks up huge quantities of heat – the water gets really quite warm --- great for swimming in, or for feeding hurricanes. The Stream hits the Americas and deflects north, flowing along the USA's eastern seaboard, still gathering heat. The GS then runs into Canada and turns eastward. It crosses the North Atlantic, headed for England and Europe. It is a huge stream of very warm water, now at high (cold) latitudes. As it crosses the subarctic northern Atlantic, the GS dumps its excess heat into the cooler atmosphere. The contrast between arctic air and warm GS waters produces the ferocious fogs and the rainy but very mild-temperature climates of England and northern Europe. The Gulf Stream is what makes England habitable. Ultimately the Stream splatters against Great Britain, and the Scandinavian and European coasts. Thereafter what is left of the Stream turns south and dissipates in the Atlantic."

Anita snorted gently: "Rebecca, I spent a year in London, and know about the fogs – and that weather is all because of the Gulf Stream?"

"Absolutely… today's nice mild British climate is the Gulf Stream's fault. More important than fog is the warm air produced by the GS - it makes England and most of northern Europe habitable for both humans and our crops – especially cereal grains from which we humans get the great majority of our calories."

"Rebecca, do we have any idea what the British climate would be like WITHOUT the Gulf Stream?"

With a grin Rebecca replied: "I thought you'd never ask! We do indeed, and that's where we're going in this discussion. But first we need a little geography, and then we can talk about the probable effects of all that ice and fresh water arriving in the North Atlantic."

Rebecca brought up a small chart of temperatures, and introduced it:

Mean air temperature at ground level, in °C

MONTH	J	F	M	A	M	J	J	A	S	O	N	D
LONDON	5	7	9	11	14	16	19	19	17	13	10	7
PETRO	-8	-7	-5	0	3	7	11	12	9	4	-2	-6

"Why is all this important? Consider London and the Russian city of Petropavlovsk (on the Kamchatka Peninsula) as "sister-cities" geographically. Both are at 52°-53°N: each abuts a major ocean and both are in nominally "oceanic" climatic regimes. Their climates should be roughly comparable, but they are not even close. Look at the data, everyone. The average monthly temperatures for Petropavlovsk are up to 14°C **colder** than London's. Fourteen degrees C is nineteen degrees F. That is an ENORMOUS difference."

She paused for effect, then carried on: "The extreme differences between the two similarly-situated cities are a direct result of the Stream's influence. In short, without the Stream to warm the continent, London (and Paris, Stockholm, Berlin, etc.) would have approximately the weather and climate of Petropavlovsk. Without the Gulf Stream, in those regions it would be impossible to grow most crops."

"The message is this – the freshwater and ice from the Greenland Ice Cap are about 100% certain to shut off the Gulf Stream. Dead stop. Probably in a very small number of weeks. We do not know if, much less when and why, the GS will come back on-line."

Anita stifled herself, emitted a muffled "Oh my god!"

"I do believe, Anita, that you just now got the message. I hope our audience does, too. We –meaning much of mankind- have a HUGE problem coming at us very fast indeed, and there's absolutely nothing we can do to change it – it's all simple physics on a planetary scale. But if the Stream does turn off, that will not be particularly unusual or unique. Geological records from sea-bottom sediments show that the GS has often shut off completely for quite variable lengths of time - from perhaps only months up to thousands of years. As I said before, and as I now repeat for emphasis, we haven't a clue as to what triggers the Stream to re-start. We do know, from sea-floor geological records, that the Stream can shut off amazingly fast - probably within a few days, certainly within a few weeks. All this is scary when you consider the human consequences of it shutting down."

Anita was both stunned and appalled. After a couple of audible gasps for breath she managed to sputter, "But Rebecca - SHUTTING OFF THE GULF STREAM? An entire, big, fast current? Can that really happen, just because of some freshwater? That sounds absolutely nutso!"

To her credit, Anita at once understood -intellectually- the implications of shutdown, but emotionally she couldn't quite believe what she'd just been told and would shortly be reporting to the world. "Rebecca" she said, "I'm at a loss here! The whole idea of the circulation of an entire OCEAN, the North Atlantic, being shut down almost instantly and perhaps permanently by such a trivial amount of water – trivial, I mean compared to a whole ocean – well, that just doesn't make sense to me."

In preparing, Rebecca had anticipated such a reaction. "Think about it this way" she said: "First of all, we're not talking the ENTIRE circulation of the North Atlantic… not even close. We're worrying about a relatively small percentage of the overall circulation… a bit of the circulation that happens to have special connections to some human activities and locations." She grinned: "In high-school, I was a bit of a tomboy, and loved to work on cars. My best friend's old Ford suddenly developed a bad habit of just STOPPING. The engine would go stone dead, at random times. It takes three things to run a car engine – air, gasoline vapor, and spark. Check – gas in the tank. Check - air available and not constricted. Check – ignition working fine. Crank till the battery dies, it won't start. Put a little raw gasoline down the carburetor and the engine fires just fine, runs for a few seconds, dies. But if you just wait a few minutes, patiently, suddenly it'll start and run normally – for ten minutes, or maybe for three weeks. Very puzzling. We couldn't figure it out. Long story short, when we finally went to change the fuel pump, we found in the gas tank a small dragonfly wing. No idea how the dickens it got there. It would slosh about in the tank doing no harm until at random it got sucked across the open, intake-end of the fuel-line inside the tank. Bingo, no gas. Dead engine. Stayed like that until the wing would float free. That wing probably weighed well under ten milligrams, and it would stop the whole darned two-ton transportation system absolutely cold."

"The point being, Anita, that many very powerful processes can be turned on and off, or otherwise controlled, by proper application of relatively trivial forces. I think it's safe to say that oceanographers in general believe that the Gulf Stream's driving mechanism can be turned off by a layer of freshwater just like what is coming. We DO know that the Stream has both wobbled back and forth a lot, and also turned off and back on again many times during, say, the last half-million years. If this Greenland freshwater DOES turn off the Stream, the Stream will most likely also turn back on again – eventually someday - but I don't believe anyone, especially including me, has a very good idea as to WHAT might cause it to re-start, much less precisely WHEN. Certainly any eventual re-start is not under human control."

"I understand that such behavior sounds unreasonable, almost ludicrous, somewhere between ridiculous and impossible, but the data on starting and stopping are quite good. Everyone is probably wondering how this can happen. Keep in mind that we're talking about a layer of freshwater on the order of tens of metres thick. That's 160 feet ... not a trivial amount of water. And it's going to FLOAT atop the North Sea's saltwater: floating is critical."

"Here's the scenario. The best way, and perhaps the only way, to shut down the Stream is to turn off its "motor" - its driving force. We're about to have a goodly layer of GIC freshwater spread across the production grounds for high-density brine, and the layer will almost certainly turn off the GS's 'engine'."

"OK, But please explain to this non-oceanographer just exactly **HOW**?" asked Anita.

Rebecca replied. "The process seems odd, but it's really not complicated. Let's walk through what I think is going to happen, and why." She took a deep breath as if to settle herself down. "The incoming low-density freshwater will float atop the North Atlantic's higher-density salty waters. As winter comes on, the upper surface of the freshwater will freeze, but that freezing WILL NOT PRODUCE the high-salt brine that comes from freezing seawater. The freshwater layer will slowly freeze from the top down, just like a freshwater lake does. Ice is an excellent thermal insulator, and the thicker it gets, the harder it becomes to cool the underlying saltwater. That cooling of saltwater is precisely what is needed to create the brine that powers the Gulf Stream. But that strong surface cooling will never reach the saltwater under the new freshwater layer."

She shrugged: "This is not just me being alarmist. The data are good, the physics and chemistry are very simple. Let's examine the chain of events again,"

Here Rebecca put up a word-slide:

'Freshwater covers the central North Atlantic' hence
'No brine production', hence
'No driving force for the 'conveyor belt' hence
'No Gulf Stream.'

"Anita, the guaranteed-predictable effects of a GS shut-off on today's human affairs are so huge it's really difficult to envision them. The **unknown and unimagined consequences** are also likely to be extremely important. Most likely, in dealing with all this, the most important and difficult problems will be things that nobody ever, EVER thought about – the results of Murphy's Law and the Law of Unexpected Consequences. At any rate, I'm sorry to be the messenger, but for much of Europe, stopping the Gulf Stream will mean an almost instantaneous change away from the present lovely soft continental climate in the direction of sub-arctic conditions – with all that such a change implies. Like, for instance, no more cereal crops in England. Et cetera."

Anita: "I want to be sure that I, and our viewers, are correct in thinking that we're discussing things that can happen RIGHT NOW – essentially today or perhaps in a few weeks – not some indefinite future date. Is that right? That we likely have only a few weeks to think about this and get ready – whatever "READY" means!?"

Rebecca said "That's right, a few weeks at best."

Anita's turn to shrug resignedly. "Rebecca, I'm a bit of a cynic when it comes to politicians and governments: I know that "NOW" is a word neither liked nor even understood by governments and politicians. But the reality sure seems to be that if and when the Gulf Stream suddenly disappears, then people in the millions will face starvation and freezing to death – and a great many of them in some very affluent societies. If shutoff comes, it is likely to be complete and to take only days or weeks. It would seem a good idea to get going NOW on plans, wouldn't you think?"

Rebecca just nodded silently.

Anita mulled this over for a bit: "So, Rebecca, is there anything at all that we can do, other than get ready for permanent, disastrous changes in climate and weather?"

Rebecca shook her head almost sorrowfully: "I don't see what that might be, other than things like looking for grain to cover the missing harvests. These events are primarily physics, and are occurring on scales where mankind is utterly unable to control anything. Look, let's be very blunt and simple. There are already over thirty thousand cubic kilometres of GCI water and ice on the move at about two MPH, and frankly, it's going to go where it damn well pleases. At the moment that destination is almost certainly atop the power generator for the Gulf Stream. We can't deflect it, we can't melt it, we can't grind it up to make it go away."

Then: "And that's not to even consider that some more of the remaining 98% might decide to emigrate to the North Atlantic. If more water joins today's input, all bets are really and truly OFF."

"So…" said Rebecca – "…basically, it seems to me there is nothing to do except try to predict, and get ready to accommodate to, the massive social and economic and political changes that will occur . Not "MAY occur", but WILL occur."

Anita took a long breath, returned to the fray: "I'm almost afraid to ask, Rebecca - what sort of social problems?"

"Oh, it's things social all tangled up with economics and politics and other aspects of life. Let's consider just one problem for the moment - as if it can be isolated from everything else, which is nonsense… but the thought experiment is a good exercise. Suppose we choose the problem of heating London. London is a good example of a very large city manifestly NOT designed for instantaneous imposition of an arctic climate for seven months per year. Neither is London designed to be readily modified to handle such a change. Think of what MUST happen socially and economically when, in the course of one year, the mean summer temperature drops from about 18°C down to 5° or 6° C - and the new midwinter mean is minus 8°C instead of plus 7°C. Where will London get the oil and gas needed to heat this huge, suddenly-arctic city?"

Anita snorted gently: "Well, Rebecca, my cynical analysis suggests 'Probably not from England's long-standing friends in the Middle East', you can bet! Not from the North Sea's oil-fields either: after all you've told us here, and looking at those satellite flow-maps, it seems pretty much inevitable that the oil-field part of the North Atlantic is going to fill solidly with ice nearly a kilometre thick. I know that most of those oil-fields are located in relatively shallow water. So I guess the ice will run aground on top of them. That'll wipe everything off the bottom – all the well-heads and pipelines, won't it? Probably even the buried stuff will be destroyed!"

Rebecca nodded: "You're right – a few feet of mud atop a pipeline is pretty poor protection from an iceberg. Now, along the lines of heating homes and businesses and so forth, it may be possible to eventually render much of London's human habitation "warm enough" – whatever that means. I think I can guarantee that the meaning today is NOT going to be the same as the meaning next year! But getting to "warm enough" may well mean spending some truly huge percentage the residents' mean income just on heat."

"Then, too, such a "social" change might have less-obvious ramifications. For instance, given the near-certain large-scale loss of jobs and the likely-to-be-astronomical cost of heat, a good many people may decide to leave for better climatic and economic conditions. The Middle Eastern refugee problem may pale when the sign of migration changes. There's a possible 'unexpected consequence' for the politicians to mull over."

Anita: "Ugh. That makes my head hurt! Maybe we should switch away from the obvious and immediate, and discuss larger-scale consequences of a shutdown. It seems obvious that effects will not be restricted to European areas overtly affected by the Gulf Stream. Will they?"

"Nope – nobody in this play gets out unscathed, I'm afraid. For discussion purposes, I'm willing to venture some opinions about larger scale effects. Which means things like weather and climate. Here, I must once again issue my repeated disclaimer, namely "I am NOT a specialist in such things!" But some of the possible consequences on those scales seem straightforward, and we oceanographers do think a lot about the interaction of the oceans with the atmosphere."

"For instance, the locations and movements of large masses of surface water of various temperatures help determine where the jet stream goes, and how strong and how variable it is. I mean LARGE bodies of water such as the North Atlantic, or the big central gyres of the Pacific. Those water-bodies help determine how far north and south the various wiggles in the jet stream go. Those wiggles move global rainfall patterns around. At this moment in global history, on average a goodly amount of rain falls on temperate-zone lands, making it possible to grow abundant crops in those places. But those patterns of rainfall can change literally overnight... perhaps the best-known example is the El Nino/La Nina cycle. The scientific community, after about 75 years of hard, expensive work, can make some useful, reasonably accurate predictions about the oceanic conditions responsible for El Nino… the predictions are not what one would call "really good" but they are at least useful."

"For a Gulf Stream turnoff we can't make good predictions. After all, modern *Homo sapiens's* entire history, not to mention the world's scientific community, hasn't included even ONE such event – the most recent "on/off/on" was at the end of the most recent ice-age, say ten to fifteen thousand years ago. But we certainly CAN examine reasonable "What if?" scenarios. Here's a good one to mull over – suppose the GS stops as I am suggesting. That will drastically change the sea-surface temperature regime of the North Atlantic, which will surely lead to a shift in the position of the jet stream--- and that shift might very well have a drastic effect – like, dumping into the Pacific most of today's North-American continental rainfall. Think about it - that water is presently being used by farmers in Jayhawk's home state of Kansas, to produce wheat and corn, most of which is exported, and very probably a significant amount of those exports go to Britain."

Rebecca wound up by saying "Wheels within wheels, effects upon effects. I think it's important to make clear to our viewers that the effects of the ice-cap's collapse on the ocean is something about which we have some firm understanding, and about which we can make a few reasonable predictions. And viewers also need to keep in mind that the effects on atmospheric circulation and rainfall patterns is much, MUCH more speculative, and not very predictable. Certainly not by ME!"

Anita muttered, "I do hope the British government will wake up and turn out to be capable of action! Then the rest of us can follow."

●●●●

The Gods of course overheard that 'woken up' comment. And chuckled loudly. Those in the know understood that there was one final huge shoe to be dropped, and looked forward gleefully to the event. But not hurriedly.

●●●●

Rebecca re-started the conversation. "Okay, let's get back to work. We've covered some oceanography and physics, now let's talk biology. The expectable biological results of having The Puddle in the North Atlantic follow from very simple biological reasoning. First of all, within a few days, the coming thick surface layer of freshwater will be a near-perfect barrier to plant growth, and therefore to oxygen generation anywhere near the surface. And 'near-surface' is where most ocean organisms live, feed, and breed. All that living stuff is absolutely oxygen-dependent."

"Consider a fish's dilemma… its food is supposed to be upstairs, on the roof. But suddenly there's a thick layer of sterile, functionally deadly water up there. Even the freshwater alone will quickly kill most marine critters: no fish is going to venture upwards past the bottom of The Puddle and into fresh-water. And even if our fish were to make that trip, it would find neither oxygen nor food.!"

"Won't algae eventually grow in The Puddle?" asked Anita.

"Not well, if at all…" replied Rebecca – "… very few marine algae could survive in the freshwater layer – if they could even get there in the first place! Where would the algal seed-stock come from? Not from the salt-water, and the freshwater, remember, is sterile and lacks nutrients."

"So what's going to happen to the ocean ecosystem, Rebecca? The ecosystem and fisheries and such?"

Rebecca answered: "Think about how the system works. Human fisheries are a reasonable place to start. Most commercial fish species live during the day at mid-depths but come up to the surface at night to feed on plankton. With The Puddle in place, first of all there's nothing up topside for fish to eat, and second, the thick layer of zero-oxygen freshwater is essentially deadly poisonous and no sane fish will ever swim up into it."

"That lack of oxygen is difficult to remedy. Without any resident algae to produce oxygen in-situ, the freshwater can get oxygen only through the dissolving of atmospheric oxygen into the surface water and then having the oxygenated waters be very, VERY slowly transported downwards by the upper ocean's inefficient mixing processes. It will certainly take years, if not decades, to restore normal both near-surface oxygen concentrations, and normal salinity. Meanwhile everything that is oxygen-dependent simply crashes."

"Anita, guessing by what we know of the Gulf Stream's activities in the past, The Puddle's anoxic condition probably won't last for any geologically significant time period… but it will quite likely last long enough to wreak havoc on mankind's Gulf-Stream-dependent activities in the ocean … primarily fisheries. The fisheries in much of the North Atlantic, and much of the rest of that ocean's ecosystem, will probably collapse in parallel."

Anita again summed things up: "So, Rebecca – my understanding of what you just outlined is pretty simple: wherever the freshwater goes, the result will be like this: (a) near-surface, there will be plenty of sunlight, but no seed-algae, no plant nutrients, and no oxygen… hence no fish-food; (b) down deeper, beneath The Puddle, there will be plant nutrients and the original near-surface saltwater algae, but no light. In both places there would be no growth of photosynthetic algae and no production of oxygen."

"Bravo, Anita – precisely correct again."

Anita grinned: "I'm almost afraid to ask this, but just how long is The Puddle going to last? What are your thoughts?"

"So - exactly how long The Puddle will last is a big unknown. Physics imposes a certain degree of "temporary permanence" on the unfortunate conditions because the mechanisms to disperse The Puddle work slowly. To get rid of The Puddle, I think there are two possible mechanisms: export and dilution. To return to "normal" surface salinity and temperature in the central North Atlantic would surely require **both** transport of freshwater, which means "ship it out of the region", **and** strong mixing, which means "dilute it down!" Some continual physical loss of fresh water from The Puddle is inevitable, due to the North Atlantic's overall surface circulation, which normally exports surface waters to the South Atlantic. We can only guess at possible rates of the processes – I have no idea yet, myself. But both the North Atlantic and our Puddle are big bodies of water, so it will probably take at least a few years to export most of the freshwater."

"So - there remains 'mixing' to discuss" said Anita.

Before moving on to discuss mixing, Rebecca issued another disclaimer: "Please, all you viewers out there, remember that the physics of mixing is NOT my field… although I do know something about it. I will probably sound a lot more certain than I feel. SO – the North Atlantic is unlikely to get rid of very much of The Puddle by mixing the freshwater with underlying saltwater. Why is this so? Because to mix fluids of different densities is really, REALLY difficult, folks. Especially on this scale - and I say that with an exclamation point! The huge difference in density between sea water and the overlying freshwater will suppress any turbulent mixing between them. Mixing requires putting energy into the system because you have to do a lot of WORK. For mixing your coffee, you provide the energy input with a spoon. But in the case of the North Atlantic you will need need a huge source of energy - something neither known nor believed to exist. Think about what mixing requires! To mix the freshwater with salt water, an enormous amount of deeper, denser water must somehow be raised against gravity, or, alternatively, the upper less-dense layer must be forcibly submerged against its own buoyancy. Both of those processes require a huge input of energy. And once the raising and submerging have been done, you need even MORE energy for doing the actual mixing. For all three processes - raising, submerging, churning - there exists, unfortunately, no mid-ocean energy source that is even remotely up to doing the job quickly. It really does look like physical export of the freshwater is what will return things to "normal". Mixing is going to be a lot less significant. Too bad."

Anita then asked, "Wait a second, Rebecca. The North Atlantic is notorious for long winters and lots of violent storms. Won't those help? Surely a big storm will do a lot of mixing… won't it?"

Rebecca replied "That's good thinking. Unfortunately, although North Atlantic storms can be spectacularly violent, their active mixing doesn't extend very deep… seldom to more than about 25 or 30 metres (and usually much less) – but The Puddle is going to be close to 50 metres thick - maybe more. Maybe a LOT more, if another increment of GIC water gets loose. So not even violent North Atlantic winter storms will mix together the different waters, down at the depth where they encounter one another. Storm turbulence just doesn't go deep enough – almost 100% of the mixing generated by winter storms will occur within the freshwater layer itself. Again, too bad: you know what people say about Mother Nature… no matter what you may think, she's LARGE and she's IN CHARGE."

"So, Anita, here's my own instant-summary about the Stream re-starting. First, for sure the Stream cannot re-start while the freshwater Puddle is in place. And there's no guarantee that removing The Puddle will somehow force a re-start. So, in order to have even a CHANCE for a re-start, The Puddle has to go away, and it looks like that will take at least a few years. In fact, even that might take longer than we expect, because for quite some time The Puddle is going to be continually replenished by melting ice."

"The available geological evidence I know of suggests this – 'Most likely YES, the Stream will eventually return… but be prepared to wait some unknowable time ranging from weeks to thousands of years, and we know not what might trigger a return.'"

Anita nodded: "Well, for our analysis and its implications, the adjectives of the day seem to be 'disappointing' and 'discouraging".

Rebecca shrugged: "Unfortunately, I have to agree. Bad news all around. Not a very upbeat diagnosis and prognosis."

16

LONDON:
CRAFTING A RESPONSE

{T_0 +2 days and onwards}
WASHINGTON D.C. and LONDON

The US East Coast's time zone is five hours behind London's. Rebecca's interview slash mini-course, complete with the London/Petropavlovsk comparisons, aired in east-coastal USA at 1900 local, running as the day's feature story. Within minutes half a dozen British scientific and diplomatic personnel located in the former North American Colonies were on the phone in near panic mode, trying (well after London's midnight) to get through to various people back in England. The British Ambassador to the USA was on the ground in DC at the moment, watwhole ched the interview thoughtfully, went pale, placed calls to Public Relations ("PR") and to the Prime Minister's Science Advisor (" SA").

The uniform initial reaction to Rebecca's clear, careful, and strong presentation was disbelief - "This is simply too outlandish to be REAL – isn't it?"

The call to PR set in motion the process of seeing whether they could control or somehow refute the story, perhaps prove that it wasn't accurate. The second call, to the SA, was caused by the Ambassador's realization of the implications if the Stream really, truly got shut down, and of the extreme difficulty of coping with the outlined instantaneous change in climate.

"London as the new Petropavlovsk, for God's sake?!!" Madam Ambassador thought to herself. "Insane! – but we better handle this story right! We mustn't be seen by the public as being asleep at the switch."

The Ambassador decided it best not to contact the Prime Minister without first having the best possible information in hand. To that end, the Ambassador insisted on being put through to the Prime Minister's SA, in London. The SA turned out to be a brilliant plasma physicist, complete with a Nobel Prize, but with little knowledge outside his personal narrow niche. He'd been chosen SA primarily because of his background – he was useful in helping to monitor the Iranian and North Korean nuclear weapons programs, not for breadth of scientific knowledge.

The British Foreign Service routinely taped all American national-level news broadcasts and it was easy to e-mail the interview file to England, whereupon the Ambassador and SA (plus his local entourage) watched the clip together, at opposite ends of a Skype real-time connection. The SA – an honest person intellectually - declared himself unqualified to either evaluate Rebecca's analysis or to advise on a course of action to be taken if Government chose to accept and act upon her analyses and predictions. He did, however, point out that her presentation had been straightforward, careful, clear, and certainly not inflammatory or self-promoting, much less hysterical. It seemed likely to be correct, especially in its larger aspects - and accuracy in details mattered little at present.

The SA understood what next needed doing, spoke mostly for the Ambassador's ears – "No matter what, we really MUST go to the Prime Minister with this. Well have to wake her up, she won't be happy, so for goodness' sake let's have all our ducks in a line when we rouse her!"

Having acknowledged his personal ignorance in the field, the SA knew they needed strong scientific support, and demanded of his on-duty aide, "Get me the Dean of Physical Oceanography at the

National Oceanographic Centre... there has to be such a person, go find out who he is, wake him up, and get him online with us NOW!"

His "official ignorance of the topic" notwithstanding, the SA could connect dots and draw conclusions, and from that make predictions. He muttered to himself, and accidentally into a live mike still connecting him and the Ambassador, "Great God Almighty, nuclear arms in the hands of the idiots running Iran and North Korea and Israel isn't enough to worry about! Now we're about to have a REAL problem, the whole damn country is going to become an arctic wasteland. Overnight, for god's sake! We'll have a permanent deep-freeze to deal with – whatever 'dealing with' means in this case! DAMN! Bombs we can probably control, at least in principle. But **this,** we've got not a snowball's chance in hell of controlling."

The Ambassador's gasp and sputter came back equally uncensored.

On his cell phone, SA punched up his top technical assistant, who answered groggily, then came instantly alert when apprised of goings-on – he'd had an undergraduate course in general oceanography a couple of decades back. Based on the boss's verbal thumbnail sketch of Rebecca's interview, he opined that the analysis sounded at least reasonable. He was tasked with checking out Rebecca's credentials and believability ASAP. SA's assistant number two was summoned, tasked with getting Rebecca on the phone NOW to talk with the SA personally.

Sixty minutes later, the sought-after Dean had been identified, located, awakened, and had received the video file and viewed it. He was shaken – the young woman had done a good job with her calm, clear, and simple analyses and explanations. Stellar, in fact, given that she was still a grad student and in BIOLOGICAL oceanography rather than physical. Good show. The thought crossed his mind that perhaps she should be contacted –eventually- about faculty positions scheduled to open up soon at his home institution? Meanwhile, Rebecca's general background and scientific credentials had been quickly checked, found very much in proper order.

Assistant #2 –astoundingly- had Rebecca on the line almost immediately. The only slowdown was due to the various participants having to figure out for themselves how to get everyone into the same electronic comm-loop, a ten minute job. Introductions all around. Immediately, discussions of Rebecca's analyses and conclusions... active debunking of over-optimistic searches for wiggle-room in things oceanographic... the 'mights' and 'might-nots'... possibilities hoped-for but not to be used for planning. Fact versus wishful or wishy-washy thinking. Rebecca's conclusions further explained, examined, discussed and concurred with by the Dean. Dismay over the likely long-term climatic consequences, with "Petropavlovsk Version Two" definitely not a welcome thought... agreement that if the Stream simply shut off overnight, "click!", the British climate's reversion to arctic or subarctic would probably take less than a long weekend.

Muttered by someone through a thick Scottish accent, "Good Lord, they don't raise cereal grains in Kamchatka. I've been there. Nasty place even at midsummer. If that's what we're in for in London or in Britain generally, starting say two weeks from right now, then we're all going to starve or perhaps just freeze to death. Best not to plan on an extensive grain harvest this year in areas dependent on the Stream's climatic influences. And nobody on this planet -except possibly the Yanks- has a year's worth of food in reserve. NOBODY! And most certainly not Britain!"

Another voice: "And if they're smart, the Yanks won't sell us much of their stockpile, because who knows how all this oceanographic stuff is going to affect the jet-stream? Just as Ms Wilson said, it could move the jet stream so that it dumps most rainfall into the ocean instead of onto Kansas or Australia or Ukraine or someplace similar, where one can grow wheat and maize. The Americans –or ANYbody!- might ultimately wind up in a different version of our British predicament."

Somebody else joshed about buying grain futures for one's own personal account: "... but you'd better do it QUICKLY, before the London and Paris exchanges open with this news on the front page."

Mister SA pulled the meeting back to order, thanked Rebecca profusely. The SA asked if she could be on five-minute standby for the next few hours, to perhaps talk directly with the Prime Minister, perhaps also to participate ex-officio in a rump meeting of various English Secretaries of State, which meeting he was sure would happen within a very few hours.

Rebecca was gob-smacked at the whole idea, got ahold of herself before her smart-aleck answer escaped ("Oh, I suppose I could be available. I doubt I'm going anywhere in that time frame, and I certainly have little better to do....") – she actually said, "Happy to help, call me absolutely any time you wish. I guarantee to be available."

Decision, at about 0130 GMT: call ahead, wake the PM, give her ten minutes to get presentable and ready for a serious conference, declare it to be "fate of the nation" level – that should wake the Old Girl, they all thought, privately.

TEN DOWNING

After Madam PM and her SA had watched the whole 'Rebecca lecture', she found herself both impressed and convinced. She asked the SA, "Can you get Ms Wilson on a phone line? I need to get a personal reading on this woman, then ask her a few questions, not about the science, but other things."

The PM went on - "Immediately, right NOW, I need a conference, face-to-face, participants will be the Secretaries of State for the Home Department; Defence; International Trade; Communities and Local Government; and Environment, Food and Rural Affairs; plus yourself and the Dean of Oceanography. I may also ask that Ms Wilson be patched into the get-together. Someone must get from her, instantly now, a confidentiality agreement."

All the named Secretaries were in-country at the moment. They were roused from their slumbers by red-phone, then picked up and driven to Ten Downing, within thirty minutes. There, the group ran Rebecca's video once more, for the late-comers' benefit.

Some minutes later, aboard Jayhawk, the duty radioman turned to the Captain with a bemused and puzzled expression, said "Excuse me, Captain, but I have someone on the line who wants to speak immediately with Dr Rebecca. Says he is the Science Advisor to Her Majesty's Government, particularly to the Prime Minister." He almost giggled, then caught himself: "Sir, this caller claims to be working in the office of the Prime Minister of England, and that the Prime Minister herself wants to talk to Doctor R. I don't think it's a prank. Shall I patch the call through to her lab?"

The Captain grinned, said "Not only YES, but HELL YES! Rebecca said something like this might happen. What a hoot for all of us but especially for her. Great way to begin a career! Put the call through!"

Two minutes later, Rebecca found herself shivering nervously and staring into the lens of her computer, running Skype. The SA wasted no time. "Ms Rebecca Wilson, meet the Prime Minister – Madam PM, this is Ms Wilson, the oceanographer whose work we have been discussing."

The PM got down to business immediately. "Ms Wilson, there are also several other people in the room for this discussion, all of them very high-level officials in my Government – at a level corresponding to Secretary level in Washington DC. We have all viewed and discussed your program and we are taking your conclusions extremely seriously. I assume you understand what a monstrous disruption of the social order your predictions lead directly to, should they eventuate? Directly and inescapably? Not to mention the chaos and expense of getting going on a response!? We should really truly hate to cry "WOLF!" on this."

Rebecca tried to remain cool: "Yes Ma'am, I do take very VERY seriously the likely consequences of any scientific information or opinion I may place into the public realm, because even the simplest of science can be, and routinely is, misinterpreted in so many ways! But please, Madam PM, remember that for me, this all began with an interview with the press, which is NOT my idea of a good time. The interview concerned the oceanic effects of the partial collapse of the Greenland Ice Cap. The program that resulted was, you may have noted, primarily me lecturing about the ocean, and discussing my simple scientific analysis of a phenomenon in which I, my research party, and the entire ship and crew were immersed. Still are immersed, for that matter."

"Ma'am, the format and content of that interview, as you have all seen, was rather "basic-educational". Certainly it was not designed or intended to get embroiled in a major policy discussion – not in Britain or anywhere else."

She paused, the PM waited a few seconds, then said with a wry grin, "Ms Wilson, you find yourself in quite a privileged position as regards cabinet-level discussions of a major national problem… after all, you are an un-vetted, unknown foreign national. I would like your word that once your presence here becomes known, you will say nothing whatever about what you see, hear, or do in all your dealings with us. No interviews or writings about our, meaning the British, end of things, without my staff's express permission. You will quickly find that it is VERY difficult indeed to say NOTHING when under pressure. Your best ally is 'I cannot discuss that.' Is that acceptable?"

"Perfectly acceptable and I agree to the conditions completely. You have my word… but won't you need some sort of written agreement?"

"I prefer your word to such a document. But the paper will follow shortly. Thank you… please proceed, sorry I interrupted you."

"Ma'am, The science in that tape is as simple and objective and straightforward as I can make it. I do not like to dance around unpleasant conclusions - that way lie all sorts of paths to failure. I stand by my conclusions – that the freshwater cap now headed into the North Atlantic is almost certainly going to stop – indefinitely- the Gulf Stream. I do not envy you or your room full of officials the task you face in dealing with the social and other consequences of a shut-down. I can HOPE that I and the others who contributed to the discussion aboard Jayhawk are in fact WRONG… but doing nothing in the face of this problem is simply betting an entire society on the slender possibility of our being wrong – actually, more accurately, of MY being wrong... my analysis therefore mea culpa. The shutdown isn't going to happen tomorrow, but might very well happen within a smallish number of weeks. The time for a floating object to travel across the North Atlantic via the Gulf Stream is only a couple of weeks."

Rebecca paused, then spoke directly to the PM. "Madam PM, Britain has fine oceanographic institutions and scientists – my entirely unsolicited recommendation is that you put them to work immediately to try to disprove my predictions – I will help in the task if you wish, and I should be delighted to be proven wrong. But for heaven's sake don't use your scientists and their efforts as a reason to delay planning for the worst. I mean no disrespect, but that would be pure idiocy, Ma'am."

Madam PM grinned, looked around her, then back at Rebecca. "Refreshing candor, Rebecca – a well-known American trait. Much appreciated by me, and in very short supply within Ten Downing, my dear. At all times, and particularly during stressful events. I could not agree more. That's what we will do."

Once Rebecca and the PM finished their conversation there were only a few minutes of questions, after which the PM said "Ms Wilson, you have been extraordinarily helpful, from your little icebound floating prison. We do hope you get loose shortly. Apparently when you do escape the intent is to have your vessel stay on-scene to do further studies of the freshwater Puddle – with yourself as chief scientist for the effort. If that is correct, we would of course especially appreciate rapid access to whatever information or interpretation you come up with. If you wouldn't mind being kept on tap for an indefinite period, I would greatly appreciate it and try not to take too much advantage of you and your time."

Rebecca replied, "Ma'am, I would like to be as much help as possible. Would it serve your interests to be at the top of my list of people to receive advance copies of any further interviews or analyses I might undertake? In which works I guarantee there will be no slightest use or disclosure of anything I learn here."

"Yes, thank you, such help would be most highly appreciated. My SA's staff will set up a communications channel."

The PM stopped, thought for several seconds, then announced, "Well, ladies and gentlemen of Her Majesty's Government, I have heard and learnt enough so that my present view is that this situation reaches the highest level of national emergency." She smiled at her retinue: "Hence you are all forgiven for waking me. Let us do two things beginning immediately: first, a detailed, thorough scientific review to be ready for public dissemination via BBC on today's five o'clock evening news. I simply MUST be able to say to the British public, definitively, either (a) "YES we have a problem let's get cracking on it", or (b) "More alarmist claptrap, have a nice dinner and sleep tight."

Then directly to Rebecca, "Thank you again for your help and offers of more. We will now sign you off for the time being, Ms Wilson, and we will begin defining specific problems and useful approaches to them. I'm sure we will contact you again shortly."

Rebecca leaned back in her chair, utterly amazed at events. She thought for a couple of minutes, wondering what else, exactly, she might contribute. Then she abruptly realized she was ravenous, and set off to fix the problem. In the mess-hall she had her first taste of persistent questions from those people curious about her dealings with the Brits – but at least these were colleagues, many of them friends, and when she explained her situation they accepted her silence.

Back at 10 Downing, after Rebecca exited the meeting, once again the video-interview was the starting point... with side commentary primarily by the Dean. When the tape finished, the entire group sat stunned and silent until the Secretary for Defence managed to say "London's climate changes to that of Petropavlovsk? Permanently? Starting next WEEK!? That surely ought to be bloody impossible! Hell and damnation, ladies and gents, we're supposed to be worrying about global WARMING, not the sudden onset of continent-sized deep freezes! Where the bleeding hell did THIS come from?"

The Dean of Oceanography was now the obvious scientific lead. He had been unable to find any holes in Rebecca's logic, and said so. Even now, at this early stage, he was convinced... the only real question was, as he put it, "How long will it take for London to go Petro?" (thereby coining a phrase soon to become a permanent fixture in the language).

The Dean told them, "As the perceptive Ms Wilson suggested, my faculty colleagues are doing an independent analysis as we sit here – I've got the best people in my organization working as fast and carefully as possible. This 'Rebecca-the-Oceanographer' clip has already run in the USA, and it will start running in Britain with the early morning BBC news in about four hours. I'm no psychologist but I KNOW that for sake of public sanity and to avoid panic, you, Madam PM, must make some sort of strong clear public statement ASAP. In dire situations such as this, the people need a leader, preferably one in full view, wearing a tall fuzzy white hat and sitting imperturbably upon a white horse, and to be told what to do or not do. I put an 11:00 deadline on my people to have a good preliminary analysis done and in this group's hands. They'll produce it, and it will come with a one paragraph introductory summary in purest layman's language. I also suggest that we get Ms Rebecca to vet it... she's going to be an important figure in all this. She seems a natural at explaining things to lay audiences, and if she is happy with what we proposes to tell the public, so will I be. But note, everybody, that I do NOT repeat NOT expect much change in her views – she's remarkably thorough and quite good... conservative in her conclusions, actually. And in fact the situation is essentially quite simple. When my folks' analysis gets here, we can use it as an outline for our immediate public response – which has to go out on the evening news, not later- and for our preliminary plans for dealing with the problem. Which we better start thinking about immediately!"

The Dean paused again, shook his head. "You know, I am now rather long in the tooth, and have memories most of you younger folks do not. Decades ago we had a formal governmental organization entitled "Civil Defense" – it was part of their job to dream up these sorts of scenarios and then consider in detail how to deal with them. Asteroid impacts, reversals of Earth's magnetic field, massive volcanic eruptions, epidemics, huge solar flares. That organization was disbanded years ago in the name of economy. Too bad. We'll just have to fake some believable veneer of preparedness."

The PM shrugged, said 'Good Lord, how I hope both Ms Wilson and you, too, Dean, eventually turn out to be all wet. However, we cannot afford a false start, nor can we brook any delay. In this room, we will act as if we're certain the worst is going to happen. We will carefully consider how frank to be with the public. Nobody makes any statement to the media until our own dominoes are in line. Say only 'We're working on it.' Or something equally controversial and informative."

She scanned the room, asked for further comments.

The Environment & Food Secretary had been silent thus far, but now he spoke up: "Madam Prime Minister, my Department has to my knowledge done zero planning, even as a paper exercise, for this possibility. So I'll wing it, ad lib and not for attribution! Heat and food are the two core civilian requirements - disregarding air and water. And disregarding the fact that the whole damned national economy is likely to collapse. But that, too, we can set aside for the moment. Winter's going to arrive pretty quickly: I think we must first tackle food and heat. 'Heat' meaning fuel- all else can in principle be taken care of later. There are some important things to think about."

He began ticking off points on his fingers. "This country depends heavily but by no means entirely on our own food crops. Most of our human-dietary calories come from wheat – plus rye for brewing and considerable maize mostly for livestock feed – no great percentage from all the other cereals combined. We'll just not even worry about things like veggies, which will be in exceeding short supply, I expect... who amongst us here knows whether or not tomatoes and spinach grow in Petropavlovsk? My guess is 'NO'."

He shrugged, forged ahead: "Latest wheat-consumption figures per year for GB are, to my best memory, about 1.6 million tonnes of domestic wheat, plus about 200 thousand tonnes imported. Barley, about 450 thousand tonnes, all domestic production. Neither we nor anyone else has a year of reserve – except for the Yanks, who probably do have it, but if so it's scattered and uncoordinated – no central accounting or authority exists over there in the Wild West. Britain ended last year, I believe, with about a year's supply of barley and perhaps two months' supply of wheat on hand. If we're going to get no crop this year, then we'll need to immediately get cracking, looking for two million tonnes of wheat at a minimum. That's tonnes per annum forever, just for us Brits, disregarding the needs of the rest of the world including Scandinavia, Central Europe, et alia. Worldwide, that should be doable but it'll raise holy hell with commodity markets and the price of wheat specifically."

The Secretary looked around the table. "IF London goes Petro as predicted, then we'll run out of wheat about 8 to 16 weeks from NOW. During mid-year we usually have a couple of weeks' worth of raw wheat in the overall domestic supply-line, but the grain coming to the country's dinner-tables right now is the tag-end of last-year's harvest coming out of the storage bins, and the pipeline is about empty, in anticipation of the harvest we now seem likely not to get. The imported part of our supply comes from America, Canada, and Australia… if we are lucky, and if we can PAY adequately, we might get them all to increase their sales to us by a factor of say five – which nets us only half what we need – half "normal" consumption/supply. A factor of five overnight is a huge stretch – we needn't dream of a larger factor. There are big "IF"s painted all over the idea. And food-demands are pretty inflexible. We'll have to go shopping for the remainder, a percentage point at a time. Ugly prospect. Just think of the inevitable price gouging."

Defence nodded, said "If the main European grain-lands are likely to get the same climate treatment as ourselves, then we can't go looking for replacement grain from the Continent. That means bringing in one hell of a lot of wheat from overseas. I assume bulk grain arrives only in ships, correct? One couldn't, say, airlift in that much raw grain? A Berlin Airlift Redux, to stagger the mind?"

The others shook their heads 'no', almost in unison. Not doable.

He continued "I wonder if the shipping capacity could be gotten on this sort of notice... assuming we could both find and pay for the grain?"

The International Trade Secretary was experienced in transoceanic shipping by sea: he responded "My guess is that if wheat supplies can be located, the bottoms for getting it here can be made available, perhaps with a bit of a premium for short notice. A.k.a. Defence's 'gouging'. Some numbers on this topic - two million tonnes is a lot, but most grain-carrying ships haul say fifty thousand tonnes at a clip, which means we'll only need about 40 shiploads per year, roughly one per week. Quite a small number of dedicated vessels could easily handle the job. Remember, consumption is spread out pretty uniformly over the entire year – we don't have to ship the whole year's supply at one go. Two megatonnes per year comes down nicely to about one 40,000-tonne shipload per week – that will hardly overload any commercial port. In fact, it won't keep fully occupied even one PIER of a big port. If we can find and buy the grain, we can arrange the transport."

The Trade Secretary shrugged – "Here's another nasty thought about shipping - we'd better keep in mind that very, VERY shortly all that Greenland water is going to arrive... and it is carrying a HUGE amount of ice with it. The whole central North Atlantic is going to be heavily dotted with monster icebergs that will take years to melt – remember the photos in Rebecca's presentation! That's going to raise pure and simple havoc with any transatlantic maritime shipping destined for Great Britain. Or France. Or Norway. Et cetera. We and the shipping industry do not want and will not risk a whole bunch of modern-day Titanic episodes. If we can find the grain, I believe it's going to take longer than one might want to get it to us. And if the currents shove all that ice over here and pack it against our coast, that'll prevent getting ships to where they can unload in Britain... might mean sending the ships into the Med, then trucking the grain to us, maybe through the Chunnel."

"Let's stay with just the grain problem for a few ticks" said the Home Department Secretary. "Consider how many trucks it takes to carry what one ship carries. Say fifty thousand tonnes in a ship, a full-size articulated lorry holds only about 20 tonnes, so that's fifty kilotonnes divided by twenty equals 250 truckloads per ship, or per week, more or less. Certainly that's feasible, with some notice and expert scheduling."

Another Secretary interrupted - "Certainly food is a top-drawer item, but don't let's go forgetting the issue of warmth. The number one requirement is quite soon going to be electrical power or hydrocarbon fuel to keep homes and businesses in operation. We're lucky our electrical power generation capability is so heavily vested in atomics – no problems with fuel or transportation in that sector."

"The first big item will be not be food, but rather heating, which is our mishmash of oil, electricity and natural gas. Let us not for a moment forget that most of our petroleum-based fuels ultimately come from the North Sea. Which source is going to go away entirely if the floe ice jams itself into the North Sea. Which it seems quite likely to do. Hell and damnation!"

"ANYHOW – for certain, demand for fuel is going to skyrocket… even if we declare a national emergency and set all thermostats to ten C and issue free double-thick Aran-wool sweaters for all. Don't forget, NONE of our structures are built to Petro standards – they're largely un-insulated and leak energy like sieves. Both my own 200-year-old house and my ten year old Government office building are anything but air- or heat-tight. I'll bet on a quadrupling of demand for oil and gas… which is one hell of an undertaking, on six weeks' notice. I have no idea whether we can do such a thing as a quadrupling, but I'll have my people go to work on it."

He grimaced: "Just as an aside, another problem to keep in mind is that a quadrupling of hydrocarbon use won't do air quality any favor."

"My view? I think the SUPPLY is probably available – again, with gouging - but it's the damned DISTRIBUTION systems that are sure to be the weak link. Just like during the Normandy invasion. And oh by the way, just to remind everyone, there really isn't such a thing as a "British national strategic reserve of gas and oil" such as the Americans have – theirs is three full year's of present-day demand squirreled away but accessible on a few hours' notice. Pays to have one's own continent, doesn't it? DAMN George III anyhow! Any rate, we probably do need to handle warmth first, then food… you can freeze to death quicker than starve to death."

He looked around the room, said "One other thing I hate to broach already, but if things go as the oceanographers and climate people predict, then to avoid civil breakdown we're certain to need - almost immediately- a fuel-and-power rationing system. And bread rationing, of course, is sure to be needed - which will really please all our constituencies. Just like the good old days during and post WW-2. The public is going to freak, and I am frankly afraid that regardless of what we do, we are all likely to be political dead meat in the next electoral go-round unless we put on a simply stellar performance on this."

The PM nodded: "Better we all think from the start about a permanent change in climate – as Ms Rebecca said, it makes no sense to deal only with this year's impending problem. And inasmuch as this change in climate will hit other countries as well, we'd better get the Foreign Office working NOW on some sort of preliminary coordinated response."

Then, unexpectedly, another Secretary piped up: "You know, for the past decade we, and others, have been dealing with this damned refugee problem --- hordes of people leaving the Middle East and other troubled areas to come here. As Ms R mentioned in passing, we may suddenly find ourselves in the inverse situation – if the climate really does go as cold as seems likely, some parts of our economy and social structure are going to implode pretty much instantaneously. Grain farmers will be 100% unemployed. All of the economy that services grain and meat agriculture will come apart – farm machinery and shipping and fertilizers, all that sort of stuff adds up to a non-trivial bit of our economy. My own feeling is that we're just entering the greatest civil emergency to date for modern western civilization."

The Prime Minister nodded, and into the silence said, musing, "Does anyone here want the task of exploring the consequences if some other significant fraction of the damned Ice Cap decides to suddenly go out to sea. The world – and we, ourselves – best consider itself fortunate that only two percent of the Cap actually wound up in the ocean. What if, say, another ten percent takes the same plunge next week?"

Nobody volunteered.

17

JAYHAWK:
PUDDLE SCIENCE

{T_0 + about 5 days}

JAYHAWK

After the press interview, and Rebecca's personal consultation with Madam PM and her staff, things calmed aboard Jayhawk. Her crew settled in for their indefinite servitude, everyone carefully studying and discussing the sat photos as they arrived: hard-copies were posted on the mess-hall walls for all to see. Ashore, worldwide, there was an intense flurry of scientific-committee-formation. The uproar did produce results in the form of plans, albeit slowly. Jayhawk was still "Johnny On The Spot" insofar as researching the phenomenon… but she was also still imprisoned, although less securely with every successive photo.

The extra time in captivity allowed for a better scientific plan – Jayhawk would use Rebecca's instrument to study two specific items. First, the mixing rate between fresh and salt waters. Second, the instrument would be used in "slow-tow" mode seeking submerged remnants of the GS: much of the strongest flow in the northern Atlantic had already disappeared from thermal view beneath the deluge of GIC freshwater and ice. Nobody knew, yet, what was going on below the freshwater, so Jayhawk's new data on currents would be invaluable regardless of results.

Lt Jonson arrived on the bridge, a good relief – meaning she'd arrived ten minutes early. As she and LCDR Smythe stood leaning on the command console, coffee mugs in hand, they happened to be looking over the bow towards the fjord's entrance. As they watched, the 140m high ice blocking the passageway heaved visibly, grumbled loudly in an entirely unfamiliar note, and began to shift.

"Holy SHIT!" said Jonson, echoed by Smythe

The obstruction began shattering in very slow motion, shedding ship-sized blocks into the water on the outer, open-ocean side of the entrance. After ten seconds or so, Jonson muttered, with a grin, "Guess this is worth calling the Captain… wouldn't you say?"

Captain Shelton was on the bridge in under a minute, during which time the whole aspect of the plug had changed. Not only had it largely shattered, but the long core segment was clearly rotating about its vertical axis: most importantly the whole mass was shifting southwards with the general current.

As they watched, the remaining core pirouetted, side-slipped, and disappeared slowly out of sight, headed south, leaving the entrance apparently unimpeded. The three officers gaped for some seconds, then finally Jonson spoke to the duty seaman beside her: "How about you charge down to comm and get us the latest satellite view of our position. Request a brand-new photo. Pronto!" He left on the errand, at the double. Jonson turned to the Captain and said, with a grin, "Sir, beg leave to report that the door to our little garage seems to have been opened."

In the library twenty minutes later, Rebecca joined the Captain and Mister Smythe in studying the brand-new imagery, with Jayhawk herself at its center. They could see at once that there was no longer any ice, grounded or stuck, either in Jayhawk's Fjord's entryway or just outside of it. In fact, the Guardian itself was gone.

"Jeez…" said Smythe, "…talk about a powerful erosive force!"

The Captain nodded, said "This is all well and good – maybe in a day or so we'll also get down to 50% coverage out there – and do so without getting bottled up in here a second time. Meanwhile, that missing Guardian bothers me. We need a good, quick acoustic survey of the bottom in the entryway. Let's put a portable sonar into a Zodiac and send it out to check that the Guardian didn't just fall sideways and block us in here forever!" He eyed Smythe, who grinned and said "Aye, aye, Captain, I'll make it so!"

The survey took an hour to get ready, and less than an hour of boat operations to accomplish. The guardian had certainly gone somewhere, but exactly where was a mystery for another time... the bottom at the centerline of the entrance was deeper than the 200m range of the battery-powered sonar. A careful study of the acoustic records produced unanimous and unreserved agreement – there was no doorway obstacle to Jayhawk's exit. The latest satellite photo of the icepack north of them, headed their way, showed upwards of thirty percent open water only twenty miles from Jayhawk, and well above fifty percent at range fifty miles. At two knots of ice-drift, Jayhawk seemed to be within about 24 hours of ice-conditions acceptable for the prison-break. If the door remained open.

The plan was to get out of the ice-pack ASAP – to go as nearly as possible straight across it into open uncluttered seas beyond, then turn south and follow the edge of the freshwater. At normal speed for maneuvering amid such heavy ice – about 4 knots – and given some decent luck with connectivity of the ever-changing open leads, it would take under 24 hours of steaming to get to the clear North Atlantic water on the eastern side of the ice and freshwater.

PRAGMATIC USES FOR
OCEANOGRAPHIC KNOWLEDGE

A little academic knowledge can sometimes be used to personal advantage. Wayne was almost certainly the only professional commodity trader in the world's grain-futures markets who also held a PhD in physical oceanography... the expertise in math and computer modeling required for that degree had opened this unexpected career for him.

He woke as usual to the TV news, main coverage of the moment being the GIC event, the report illustrated with brand-new satellite photos of the ice-packs moving towards the tip of Greenland. He stared, quickly dug out and dusted off his ocean-current atlas. He fully and instantly understood the implications of that huge puddle of freshwater headed for the central North Atlantic.

The Greenland freshwater was going to shut off the Gulf Stream.

England and much of the grain-producing core of Europe were probably never again (at least, in the foreseeable future) going to produce a crop of cereal-grains – quite likely beginning with the crop just now growing towards harvest All other effects aside, that change was going to hugely increase the price of grains, globally.

He sat down at his home computer, got on line in his private accounts and activated their "buy on margin" features, sold everything he owned not connected with the upcoming changes, tapped his unused home equity line of credit to the max, and began to buy grain futures, worldwide, at the absolute outer limit of his finances.

Wayne was correct: within a very few hours, other traders began to realize, mostly from rumors, what was going to happen. Futures prices doubled, then doubled twice more in the space of a very few days. Within the next six months, Wayne became not quite a billionaire, but a millionaire hundreds of times over.

18

JAYHAWK:
FREEDOM + SOME DISMAL SCIENCE

{T_0 + 2.5 days}

• • •

The Gods at their gaming tables were, for the most part, either amused at or impressed by the humans' escaping in Jayhawk. ("You mean, escape THUS FAR!" some players reminded others.) There were even a few arguments about the propriety of these games - it had, after all, been aeons since the last of the great at-sea interactions between Gods and Humans! Those who had felt humans incapable even of self-preservation had of course bet against the idea of any directly-involved humans surviving to this point – chagrined, they paid up and settled back to watch the next few acts, perhaps to place new wagers. But even the grumblers had to agree that Jayhawk had avoided destruction quite neatly – and via a rather classic seagoing maneuver involving high-risk improvisation at very high speed, the stuff of eventual legends! So perhaps there was something of themselves embedded in the humans aboard the vessel? – a most intriguing, even disturbing, thought indeed! Yet it was a serious possibility, especially given various gods of both genders having an ancient propensity for slumming with humans.

Credit and cash having been given where clearly due, nevertheless there were still those players who felt that certain game parameters and conditions had been set too loosely, making things overly easy for the humans – in particular a few players fulminated against the rate at which the ice-floes were clearing out of Jayhawk's escape route. But there would be no changes: the ruling quartet held that the original ground rules were to be strictly observed. Only the simplest 'natural' physical and social forces were allowed to act within the scenario – no special forcing or supernatural interferences by any god.

But then, no further forcing was needed to produce interesting, unexpected results – for example, the fracas over in the Indian subcontinent had been an unexpected, quite amusing bonus for all – occasioning quite a bit of comment and significant hilarity, plus further rounds of ad-hoc unplanned side-betting on various possible continuation scenarios.

● ● ●

{T_0 + 4 days}

Certainly, nobody on board complained when the massive ice floe blocking the entrance to Jayhawk's Fjord disassembled before the bridge-watch's eyes, crumbling as it disappeared from view. The pieces, many bigger than Jayhawk, were quickly flushed away from the entrance and out of view by the southward current, augmented by the ongoing freshwater flow.

The Captain called HQ again, discussed events with the mucky-mucks, then called once more for new satellite photos of the area. They showed that the ice-floe coverage just outside the entryway was definitely below 50% - meeting the criterion for Captain Shelton at his discretion to take Jayhawk out. Better yet, a few miles north, floe-coverage was down to 40%, and dropping steadily.

In consultation with officers on the bridge, studying a 48-hour series of sequential sat-photos of their area, the Captain said "Looks to me like the major floes retain their positions relative to one another pretty well – at least, they don't seem to be bashing together much. Some of these big leads have maintained their integrity for almost a full day while the whole pack drifted along."

Murmurs of agreement. No churning of the floe field meant a much, much better margin of safety.

On the most recent photo, the Captain tapped the outer edge of the floe field, then said "Looks like 35 to 40 miles across, from our present position to being fully clear of the big stuff ... not a bad distance, and if the individual floes hold their relative positions as they seem to be doing, well, it shouldn't take more than about ten hours to get across. Maybe twenty. At any rate, the big leads here are a comfortable size for Jayhawk, plenty wide enough to pull a 180 if necessary."

He turned to the other officers on the bridge: "How about all of you together take these latest photos and grease-pencil a tentative course through the immediate configuration of floes. Then we'll watch the pack for a couple of hours to see how floe movement affects the path you laid out. Order up 'before-and-after' pictures." He got nods all around.

"Agreed, then," said the Captain. "You folks and the Exec bang heads together and lay us out a track that looks likely to work, then make a copy for Meadowlark to use. While we fart around studying pictures, and champing at the bit, I want a low-altitude Meadowlark survey of your idea of a route. We need to know how much small stuff is hiding between the big floes. As usual, we can expect that dodging the smaller stuff is what's going to take time and attention. If we can't go fast enough, then the time for us to get across may increase to match the rate at which the big floes grind together, and that AIN'T what we need to get involved with. We want Jayhawk to be the high-speed-maneuvering part of this evolution."

He grinned at the group: "Hey! Hot damn, we may get home for Christmas. Let's be sure we do this right – anyone got more ideas or concerns that need discussing? Speak up now if so!"

Head-shakings all around. Nothing overly complex here, just a simple maneuvering exercise in an unusually complex ice-field. Shortly it would be business as usual again. No sweat.

"XO, get on the horn and let the ship know what's up and that there's a strong possibility of Jayhawk making a break for it about four hours from now. We'll do the trip at GQ, until we're clear of the floes. Launch a Meadowlark asap to do the preliminary scouting... we've got loads of full daylight left, and even in the so-called dark of midnight we can see what's going on using the night-vision gear. Good idea for us to leave here ASAP." He shrugged, "Speed is of the essence, Ladies and Gentlemen... even if it turns out that from the photos we can predict the next few hours' ice-movement pretty well, we really REALLY don't want to have the Gods get pissed and re-block the entrance or whack us with some other amusing problem!"

The chopper left about twenty minutes later, provided with the officers' hand-drawn best-guesstimate as to a safe course for Jayhawk. One hour later, Meadowlark-2 was back aboard. One of her passengers had been tasked with making a nonstop video of the marked route. Two other passengers were widely experienced in exactly this maneuver – scouting routes for the Mother Ship through extensive fields of floes mixed with smaller ice - bergy-bits and the like.

Consensus aboard Meadowlark, when she finally found herself over ice-free water nearly forty nautical miles from the ship, was that there should be no problem maneuvering Jayhawk through the wide leads, of which there were a plethora. There was also plenty of smaller ice in those leads, but nowhere did they observe enough big chunks to cause the ship a problem getting through. In fact, the two scouts agreed, the ship should be able to make a pretty steady three to four knots most of the way. They reported so via radio, took a leisurely second look enroute home, which didn't change their opinion.

The ship's company was invited to the library for a showing of Meadowlark's video, speeded up for brevity's sake. At the end of the video Captain Shelton addressed the crowd: "Unless someone has a new issue to raise, either pro or con, we are going to make our exit in about twenty minutes. Fifteen minutes from now we're going to GQ. We will stay at GQ until we reach ice-free water on the other side. Mother Nature is still in charge when she chooses, and if we can help it, we don't need to give her any excuse or chance to do us more mischief. I'm going to call CGHQ to let them know what's going on, see if we have any special new orders. Jump to it, folks."

Jayhawk launched her second Meadowlark just before getting underway – the chopper's duty was to report on the exact conditions into which Jayhawk would be literally poking her nose. Plus sending back live video of what she was seeing.

Final exit plans were cleared with CGHQ, and when GQ was sounded there was no scurrying about – all hands were already at their duty stations.

ESCAPE

From Meadowlark's station at an altitude of 500m, her video feed showed that this instant the fjord entrance faced into a substantial lead whose long axis was functionally an extension of the fjord – a straight-ahead no-brainer exit, into perhaps three miles of good-sized lead only lightly sprinkled with small ice, all easily dodged or shoved aside. Meadowlark's crew were aces at this part of the game – they fed the bridge a continuous stream of information on location and size of nearly-submerged bits, and other details in anticipation of necessary major zigs and zags. Overall, the original grease-penciled course was holding good. All the ice, of whatever size, was entrained in the same body of moving freshwater atop the current and had been so for a couple of days by the time it got to Jayhawk's location... well before the field got to Jayhawk, most of the clanging and banging-together of big floes had been accomplished, thus reducing chaos to reasonable order. For all practical purposes, the important stuff – the big floes - was sailing along southward at two knots in very seamanly fashion, doing a fine job of maintaining relative positions. It turned out that Jayhawk could dodge and wriggle at will, with no imminent danger of being either trapped or squashed.

After an hour at two knots without incident, Captain Shelton ran Jayhawk up to four – a good clip but still a speed at which she could readily take the shock of running squarely into fairly massive bergy-bits. Meadowlarks #1 and #2 alternated three-hour flights, and collectively kept the ship from testing her design limits – she had no unexpected encounters with major ice.

Shortly before dawn, she made a final turn between two floes each several square kilometres in area, and emerged quite abruptly into completely ice-free water. That brought a spontaneous ship-wide cheer as word was passed and the ship was secured from GQ. Congratulations were extended by HQ: the entire adventure had not actually put Jayhawk more than two or three days off-schedule… now it was time to turn to oceanographic studies.

{T_0 + 4 days and onwards}

As soon as the ship was clear of the ice, Rebecca's team went to work. For measuring salinity and temperature and chlorophyll they had but a single usable instrument – namely, her 'football'. Getting physical water samples for microscopic or chemical analysis was also extremely important, but it was something her football could neither do nor be modified to do.

For science, the ship provided space, power and water, bunks and grub. But as is the case on almost all research vessels, Jayhawk's scientific lab was more virtual than actual. Each scientific party brought its own gear for measuring whatever that particular group needed to measure. The lab itself owned a few genuinely antique old-fashioned (every one of them over 100 years old, but perfectly usable) mechanically-actuated, solid-brass Nansen bottles for getting physical samples of uncontaminated seawater from any depth you had enough wire to reach. Rebecca was lucky – thanks to her tradition-oriented major professor, unlike almost all modern oceanographers she knew how to use the Nansens – hooray for the antiques! (Maybe she rated the 'Ma'am' after all?!) That oddball ability gave her the ability to collect physical samples. She was doubly lucky, because the wet-lab had some other random supplies – cast-offs from previous occupants - including several 24-count cases of glass bottles appropriate to keeping actual physical water samples uncontaminated until they could be gotten back to a respectable chem-lab ashore.

Then there was Rebecca's football – a fine example of a device NOT designed for extensive, continuous yo-yoing up and down, and also most definitely NOT intended to be towed alongside or behind a ship. Of course, Murphy being what/who s/he is, the data needed were (a) lots of vertical profiles, a.k.a. yo-yo sampling, and (b) lots of areal coverage, a.k.a. continuous sampling using a towed device.

Rebecca had long ago shrugged off concerns about the gear and research program, preferring to think along the lines of opportunity giving with one hand and taking away with the other. Design considerations aside, Rebecca had used the football over the sides of several ships, and had seen it survive accidental towing at two or three knots without damage. Importantly, they could not afford to damage or lose the football – it was really the be-all and in-all of the new, critical observational program. Carefully deployed and managed, it could be made to serve. And towing at three knots would do nicely for the physical work

{T_0 + 4-6 days}

Jayhawk spent the next two days surveying the edges of the freshwater, seeking data that would tell whether or not The Puddle was going to dissipate by mixing at the interface between fresh and salt. Jayhawk would steam to a position between two widely-spaced floes, then sit in place whilst the football was yo-yoed through the interface between fresh and salt waters, some 50m or so beneath the keel. The hope was that perhaps the patch would erode from below more quickly than rough-guess theories were suggesting (namely decades). The turbulent mixing they hoped to find would be a stirring of colder fresh water together with warmer seawater at the interface. If there was active mixing going on, the instrument should find very thin "inversion layers" where low-density freshwater was overlain (instead of underlain) by high-density saltier water – such inversions are always found wherever waters of different densities are mixing.

Unfortunately for both Rebecca's team and a small army of help ashore, analyze those data as they might, after over two days of yo-yoing came the unwelcome conclusion that there was no significant mixing taking place. The Puddle was going to have to be washed out and dissipated by the normal major circulation of the North Atlantic surface currents. Meanwhile, as that conclusion was being reached unanimously, The Puddle, with its massive flotilla of icebergs and floes, continued its steady progress north and eastward.

As Rebecca and crew were collecting their dismal "no-mixing' data, the federal government's suite of moored-in-place instruments on the US continental shelf was steadily reporting back. So far, the Gulf Stream had not reacted significantly to the situation… it was still wobbling about and pounding along at its usual four to six knots. But The Puddle had not yet covered the central core of the North Atlantic, where most of the driving forces are generated.

Speculation ran rampant as to when, and how fast, the system might shut down once the freshwater was in place – in fact, in an odd way, the Mortals' speculation rather resembled that going on in the Gods' game. Less mead was involved, however.

Amongst the humans, arguments for rapid shutdown were strong – it takes a huge amount of energy input to drive the system, and it seemed clear that after the driving force was turned off frictional forces should bring the current to a screeching halt in at most days, perhaps even hours. Counter-arguments most commonly invoked inertia… the enormous mass of rapidly-moving water that is the Gulf Stream should have so much inertia that slowdown to near zero might take as long as a year or even two. But regardless of differences of opinion as to speed of the process, all parties now agreed on shutdown being pretty much inevitable.

The British Isles and Scandinavia and much of western Europe didn't like that conclusion – but each country had its own resident oceanographers and meteorologists. Those folks had unrestricted access to all the available data, including satellite observations and Jayhawk's measurements. Essentially all analyses agreed on the upcoming death (perhaps 'hibernation'?) of the Stream.

The results of e-tech James' initial thumbnail analysis – the taste-test - had been correct. Nowhere within the freshwater itself had the football detected any trace of chlorophyll or oxygen – or plant nutrients either. Samples from the "ordinary North Atlantic saltwater" outside The Puddle contained perfectly normal quantities of the usual species of microscopic plants.

The marine-biological side of the problem was that The Puddle seemed almost certain to spread to cover well over 75% of the North Atlantic surface. The sterile freshwater had no indigenous photosynthetic species of its own, and any local salt-water species which managed to invade the freshwater (say, by being caught up in the mixing processes) would die immediately due to the huge change in salinity. In saline water lying below The Puddle, the region's normal, original populations of photosynthetic species were present, but at concentrations just barely detectable.

Under the microscope, the normally-green cells of algae and other photosynthetic organisms looked bleached and profoundly unhealthy. The freshwater was absorbing almost all incoming sunlight before it could reach the normal photosynthesizers, who were adapted to bright surface-sunlight. Photosynthetic organisms, the base of the entire food-chain, were starving due to energy deprivation, and doing so surprisingly quickly. Implications for both nearshore and oceanic fisheries were as dire as the predictions for grain crops ashore.

{T_0 + ~6 days}

For three days now, Jayhawk had worked 24-hour shifts studying The Puddle, documenting its depth, its lack of biology, its effects on the ecosystem being overlain by the freshwater. Documenting and mapping the lack of sufficient mixing to help in eliminating the fresh layer. Jayhawk had held center stage for a week, had done the basic work, had used every asset on board – the football was performing flawlessly, but the data were endless nearly-identical repeats.

A discussion had begun with CGHQ and the onshore scientific community as to just how far to push Jayhawk's efforts – keep her on station, or bring her home? Consensus: Jayhawk should be freed to return home for a personnel swap and re-provisioning. While those discussions were under way, The Puddle flowed on across the North Atlantic.

19

SHUTDOWN

{T_0 + 7 days}

Through Day 6, the American flock of moored current meters continued to show 'business as usual' Gulf-Stream-like flows. Then, on Day 7, with astonishing speed, the Stream vanished from the data of every current-measuring device. It did so over distances of thousands of kilometres. The onset of braking took a mere four to six hours, occurring in a seemingly-coordinated way that left all observers nearly stupefied. It resembled an oceanic analog of Einstein's "spooky action at a distance".

{T_0 + 20 days}

Meanwhile, even without the Stream, the North Atlantic's underlying clockwise circulation continued to drive The Puddle and its contained ice towards Great Britain. By Day 20, the GIC ice had begun arriving off the coasts of England. There the front-running floes grounded and the mass coming along behind them piled up in a slow-motion version of an icy train wreck.

{T_0 + 23 days}

By Day 23 the mashed-together ice floes had built in the Channel a perfect "plug" of ice. The plug extended from the coasts of England and France out many kilometres westward... and was steadily growing as more ice arrived.

All transoceanic shipping into and out of regions north of southernmost England was utterly stymied; only short-haul coastal shipping was possible – and that only northwards from England. This would continue indefinitely – until the ice melted of its own accord – no human icebreaking was even remotely feasible.

By Day 25 there were tens of kilometres of ice floes jammed against and atop one another, forming an impenetrable barricade which completely blocked access to Britain by sea – if the UK's needed grain could be found, it would have to be shipped into the Mediterranean via bulk carrier, then transshipped via truck or rail to England through the Chunnel. Logistics were going to be even worse than they at first had seemed... the same route would have to be used to bring grain into the northern countries, requiring yet another transfer and transportation plan to be invented and implemented quickly.

{T_0 + 24 days}

Having done all that they could, Jayhawk and crew and scientists arrived home to a reasonably thunderous welcome – at least, considerable noisy fireworks ashore and a lot of hooting from all the ships' horns in the Port of Seattle. The predictable press interviews took place, giving everyone concerned their mandated fifteen minutes or so of personal fame and glory. Books and articles aplenty were quickly produced.

And on her first night home, Rebecca received a phone-call from the Dean of Ocean Sciences in England, expressing his personal admiration and his government's gratitude for her efforts and cooperation. He had done his homework, knew she was finishing her PhD in a very few months. After some British pleasantries, he finally wondered, politely and aloud, whether she could possibly commit to a short series of lectures a few weeks hence, at his institution, detailing the "Jayhawk experience" and the resulting science? There was a significant group of faculty and students who would be most interested to meet with her. Perhaps they might, all together, discuss her career plans whilst she was on-campus?

Section Four – Antarctica – long-term consequences.

20

THE TIGER IN THE BAMBOO
a.k.a.
The Real Problem Emerges -
ANTARCTICA

(All else is commentary!)

• • •

The Gods run on a different time-system from humans, being able to slow or accelerate their personal "temporal viewpoints" at will. As a species, this we mere mortal humans cannot yet do. Only a few rare, highly educated and trained varieties of human can shift their temporal perspective to comfortably handle long stretches of time, and thereby be able to consider the progress of processes that operate on such scales – ecologists, astronomers, and geologists come to mind and give hope. The Gods had done their initial betting on very human scales and short-term possibilities and events - scales of minutes (sometimes seconds) to a few days. Now, they shifted to a somewhat longer scale, say fifty to five hundred years – a scale falling well short of the genuinely geolog-

ical (much less cosmological) but still a shift that enabled contemplation of other consequences of the GIC's demise. Not only were the time and space scales of the action now greater, but so was the importance of mid-scale consequences. Betting resumed, with enthusiasm, as the next scenario was unrolled.

• • •

When a mere 2% of the GIC's total freshwater had run off into the ocean, it had raised global MSL only a paltry-seeming 14 cm. Such a small increment – so apparently insignificant, what with the mean depth of Earth's oceans running about 4000 m. A change of one part in 28000, about 0.004 of one %. One could be forgiven for feeling that this change was trivial. But it invoked the twin laws of Murphy and Unexpected Consequences.

Most of Earth's fresh-water supply is stored in two places – underground (about 30%) and in ice (glaciers and ice caps, ~68%). The remaining 2% is liquid on the surface. By far Earth's largest reservoir of freshwater is the Antarctic Ice Cap (AIC). The AIC is huge, covering essentially the entire continent – an area of about 14 million kilometres^2, holding ~27 million kilometres^3 of ice – about 60% of the world's total freshwater.

Antarctic ice comes in three major categories: (a) the thick, stable ice cap covering most of the interior of the continent several kilometres deep; (b) abundant glaciers that flow seaward from the high-altitude central uplands, eventually discharging into the coastal ocean; and (c) large sheets of glacial ice that have finally reached the ocean but remain attached to the continental margin. Those oceanic sheets are thick and extensive: today's several million kilometres^2 of sheet-ice is shrinking rapidly. That number has varied wildly over geological time scales, including scales short enough to be perceived and understood by modern man – and to have played powerful roles in *H. sapiens*'s evolution.

The oceanic ice sheets consist of two parts: (1) the portion nearest the shore, still firmly connected to the feeder-glaciers and NOT free-floating, but rather grounded on the shallow, shelf-like bottom;

and (2) the ice offshore which is still attached to the grounded portion, but is actually free-floating.

The two parts of total sheet-ice affect sea-level quite differently. The free-floating portion has already made its contribution to raising sea level – if one melted all of Earth's free-floating ice, sea level would not change... that ice has already displaced as much water as it contains. However, most ice which is nominally "in the ocean" but still attached to the bottom would, if melted, provide a genuinely new addition to the world's oceans, and would raise sea level.

Most Antarctic ice is located in the central highlands' AIC. That ice is at high altitude and wants to flow downhill. But that downhill slide is slowed or even prevented, largely by the grounded part of the coastal oceanic ice sheets, which collectively act as plugs (or 'flow regulators'). They help to prevent the AIC from flowing freely down to the sea. In short, those sheets' resistance is crucial in checking the seaward-going propensities of the AIC... i.e., in keeping the AIC in place. Because MOST Antarctic freshwater is in the form of ice, and because MOST of it is ashore (not floating), any discharge of AIC-ice into the coastal ocean will raise sea level.

In the entire Antarctic "ice-system", the ice most fragile and most susceptible to melting due to climate change is the oceanic sheet-ice... the flow regulators. Here, yet again, tiny inputs can have enormous effects.

There is enough ice in Antarctica to raise world sea level 58 metres were it all to melt... that is 193 feet. Several entire American states lie below an altitude of 58m. Melt the entire AIC and that would turn those states into not just deep salt-water swamps, but actual sea-floor. For instance, consider the mean height above sea level of these states: Delaware 60 ft (18m); Florida 100 ft (30m); Louisiana 100 ft (30m); Rhode Island 200 ft (60m); New Jersey 250 ft (75m).

For us humans, dramatic effects from changing sea level via melting Antarctic ice do not require BIG changes of sea level. Given a sea-level rise of well under 10 m, most of Manhattan will be under water as would be all of the Netherlands. Et cetera. That is not a pretty thought either socially or economically. In short, one need not melt the entire AIC (or even very much of it) to see significant effects on human affairs.

Melting "ten-metres of ocean change" worth of water from the AIC can happen in a remarkably short period... perhaps requiring only a single human lifespan, but certainly it can happen over two or three of them. Warming of the atmosphere is the only thing we know that can provide the energy for that much melting – and humanity is well on its way to warming the planet sufficiently to cause such melting.

How would it work? Via the laws stating "Small inputs into complex systems may, and often do, create major effects, often unexpected."

The key factor is the existence of grounded and floating parts of the Antarctic ice sheets. As the ice moves offshore under pressure from the AIC and its glaciers, there comes a point where the bottom has dropped far enough so that the ice becomes ungrounded--- begins to actually float. Call this location, the line where the sheet-ice becomes ungrounded, the "Hinge Line" – for reasons to be explained.

Anything that is floating in the ocean goes up and down several times per day, with the tides. This vertical seesaw effect happens to the floating ice, but not to the grounded ice. The floating ice is going up and down outside (= seaward of) the Hinge Line, and not doing so behind (= shoreward of) the Hinge Line. The tides are flexing the ice right at the Hinge Line twice per day... and like repeated bending of a bit of steel, the ice-sheet will eventually break at the Hinge Line. The first result is the calving of Delaware-sized ice floes, as shown often in the popular press. Those floes, as mentioned earlier, do not raise sea level any further when they break free and melt. What DOES happen is some significant decrease in the flow restriction to which that now-gone ice was contributing.

As a result of decreased restriction, the flow rate of glaciers and AIC materials speeds up: those materials DO add to sea-level rise. And they are now coming into the ocean much faster than before. This is augmented by the sensitivity of the coastal (grounded) ice to melting due to global warming. As the grounded ice thins, its resistance to AIC movement likewise decreases – it is steadily losing effectiveness as a flow restrictor.

Once all the flow-restricting sheet ice –of both types- is gone, then every bit of ice (or meltwater) that reaches the coast will add to sea-level rise.

"Small effects can yield large consequences" is a truism. It just so happens that with all the decades of slow (by human standards) changes in the Antarctic ice system, the attachment of floating to grounded ice (at the Hinge Lines) is reaching a critical point, where very little added tidal action is needed to cause major effects. Adding "a mere 14 cm" of water to the ocean will raise the floating portion that much higher – whilst the grounded ice remains unaffected.

More flotation = more flexure = faster removal of both floating and grounded sheet-ice. Which means faster delivery of above-ground ice into the ocean. Which raises sea-level.

The almost inevitable loss of all coastal ice sheets from Antarctica ("almost" being one factor on which some of the Gods are basing their bets) means that sea-level will rise considerably – how much and how fast are presently unknown. But over a couple of human life-spans, that rise will not be seen as "slow".

Within merely decades, mankind will be faced with the planet-wide loss of much, perhaps most, of the coastal-plains space for living and farming and the loss of much of the enormous man-made infrastructure down at or near sea-level.

Surely there is money to be made as a result – if one can operate in the appropriate time-frame, that is.

As most certainly the Gods at their little party can do.

21

AFTERWORD: LONDON

{T_0 + 40 days}

The first snowfall of the season arrived six weeks earlier than normal. And the initial snowfall was twenty times deeper than normal.

{T_0 + 42 days}

By the end of those six weeks, it had become obvious that the re-structured temperature regime of the North Atlantic's surface was strongly affecting the location, speed and altitude of the jet stream. It was wobbling about with uncharacteristic vigor and near-randomness, driving to distraction such diverse groups as climatologists, grain-futures dealers, and airlines. And farmers.

Existing global models of patterns of both precipitation and temperature were far out of their depth, gave results as confusing as the jet stream's behavior, which verged on the chaotic. How long it would take for a new northern hemisphere pattern to emerge –with its new variabilities- and settle down to some semblance of predictability was anyone's guess.

And meanwhile, rainfall patterns had begun an ominous shift that boded only ill. The major areas of terrestrial rainfall, which support most of the world's cereal grain production, seemed to be rapidly getting drier. Across the northern hemisphere, unusual amounts of rain were being reported by offshore weather-buoys. It certainly remained to be seen what those changes would do, for example, to the flow and timing of major events in Gangaji and other rivers.

{T_0 + 53 days}

• • •

Loki, ancient Norse God of discord and evil, the cosmic prankster, was more than a little upset at not being invited to the Gods' dice-game. He took umbrage at the slight, and decided he was due a bit of revenge. He reasoned that inasmuch as he was without an invitation, the rules of the competition could not be expected to apply to him. Those inapplicables included the original commandment "No messing with the physics of the system!" He decided to toss into the works a bit of quirky behavior on the part of ocean and ice – behavior sure to render difficult the calculation of effective odds in the Gods' grand betting-scheme. His clever idea for gaining revenge required merely a tweak of system physics, easily and secretly accomplished. He could then enjoy the chaos, arguments, perhaps even combats that would ensue. Both among the Gods, and amongst the human populations.

• • •

Shortly, much to the surprise of both humans and Gods, a few seemingly-random floes of GIC-ice drifting in the overall North Atlantic surface currents abruptly turned south, against all precedent. It was quickly obvious that those floes –some of them more than ten kilometres across- were headed for the Strait of Gibraltar. There, the massive surface flow of water from the North Atlantic into the Mediterranean (to replace water evaporated from it) helped suck the floes into the Strait itself... where they grounded most satisfactorily on the very shallow lip between Atlantic and Med, blocking all water-flow through the Strait. A different form of ice-dam.

Normal strong evaporation continued unabated over the Med's surface – evaporation at a rate of over a metre per year. Evaporation without any replenishment through the ice-dammed Strait.

Sea-level in the Med began dropping at once, causing long-established shorelines to retreat seaward quite rapidly. Within a few months a great many coastal installations like piers and docks and ramps were useless. And the Med's water was getting saltier fast, to the dismay of both fish species and human fisheries.

Of course, seaborne trade suffered - for the life of the plugging GIC floes, they would make completely impossible any seaborne trade between Atlantic and the Med. This blockage caused severe international panic, because it mean that there was NO readily accessible path by sea from the rest of the world to Europe... suddenly, Britain's necessary grain supplies could neither be brought in directly to Britain, nor taken into the Med for transshipment, except via the enormously overloaded Suez Canal. A great many proposals were being advanced, addressing the problem of clearing the Strait. Suggestions ranged from using nuclear weapons to shatter the floes, to gigantic (purely imaginary) tugboats.

But there was no clever, obvious, or even feasible solution.

Loki smiled to himself, almost a smirk... patted himself on the back for a revenge well executed, and reached for his mead-horn.

{T_0 + 75 days}

The "Times of London" established a small black-bordered box on page two, displaying (1) yesterday's number of in-home deaths due to exposure to cold (island-wide) and (2) the estimated cost-to-date (again island-wide) of replacing cold-damaged water piping... a number already far into the billions of pounds.

22

LOKI'S FINAL SHOE DROPS

Loki's little rules-change had resulted primarily in problems that occurred quickly and were (to the gods, at any rate) merely amusing. But he had buried in his modifications a hidden time-delayed grand finale.

{T_0 + 15 months}

Within weeks after the GIC's misbehavior, the most-affected countries had set up a cooperative international monitoring team to watch the remaining 98% of the GIC. The 'GIC Team' had been given carte-blanche in money, personnel, and equipment. In record time, they had installed on the Cap's rugged new surface a dense network of simple transponders. Using satellites and GPS, they monitored the shape and movements (if any) of the reconfigured Cap. Measurements accurate to a millimetre over distances of hundreds of kilometres were routine.

The net had enabled them to watch the settling and compaction of the slumped materials: in ten months of detailed analysis, they had seen lots of slow-motion, island-wide shivering and shaking as the settling went on… but they hadn't seen any lateral slippage. It seemed as if the GIC regarded what the GIC Team had named "The 2% Event" as a superficial adjustment, having no need for more sliding sideways.

That, of course, was wonderful news for mankind – the 2% Event's fourteen cm of sea-level change had been quite sufficient.

Team Leader was Henry, a young British geologist who was extremely good at extracting information from complex data, such as the net's.

Henry had taken a week off - his first such for over a year. Returning, he went to his office and brought up the images taken over that week – one image every two hours. Everyone –including Henry himself- agreed that this intense coverage was overkill, but nobody fought the program.

He goggled at the images that came up on his monitor… where for months on end there had been no lateral changes, no lateral motion, now there was a real difference between the first and last pictures. Everywhere on the Cap, across the entire net, the entire island, there was displacement of transponders. In every case, no exceptions, the movements were towards the southwest.

Nervously, he had the system calculate and plot the speed of the movement, based on sequential photos.

Total displacements were in the tens of metres. Per week. That wasn't good, not at all. But far, far worse was the change between successive photos of position of individual transponders.

Those distances were, in every instance, getting larger.

Somehow, the restructured GIC was moving as a block.

And the speed of movement was accelerating.

The Beast was waking up.

And it had nowhere to go except into the sea.

With visions in his head of seven metres of new ocean arriving in Holland and Manhattan, Henry reached for the phone to warn the world.

So It Goes

—Kurt Vonnegut,
Slaughterhouse Five

www.ingramcontent.com/pod-product-compliance
Lightning Source LLC
Chambersburg PA
CBHW030918180526
45163CB00002B/378